Communications in Computer and Information Science 602

Commenced Publication in 2007
Founding and Former Series Editors:
Alfredo Cuzzocrea, Dominik Ślęzak, and Xiaokang Yang

More information about this series at http://www.springer.com/series/7899

Jian Cao · Xiao Liu
Kaijun Ren (Eds.)

Process-Aware Systems

Second International Workshop, PAS 2015
Hangzhou, China, October 30, 2015
Revised Selected Papers

 Springer

Editors
Jian Cao
Department of Computer Science
  and Engineering
Shanghai Jiaotong University
Shanghai
China

Xiao Liu
Deakin University
Melbourne, VIC
Australia

Kaijun Ren
National University of Defense Technology
Changsha
China

ISSN 1865-0929 ISSN 1865-0937 (electronic)
Communications in Computer and Information Science
ISBN 978-981-10-1018-7 ISBN 978-981-10-1019-4 (eBook)
DOI 10.1007/978-981-10-1019-4

Library of Congress Control Number: 2016936975

Printed on acid-free paper

This Springer imprint is published by Springer Nature
The registered company is Springer Science+Business Media Singapore Pte Ltd.

Preface

This volume collects the proceedings of the Second International Workshop on Process-Aware Systems (PAS2015) held in Hangzhou, China, on October 30, 2015, co-located with the 5th China Conference on Business Process Management (China BPM 2015). Following the great success of PAS 2014, PAS 2015 was a cross-area international workshop on process-aware systems aiming at bringing together researchers, developers, users, and practitioners interested in process management, and formally exploring various potentials of process management and process thinking in the era of cloud computing and big data from the perspective of both academia and industry.

As the second edition in this workshop series, PAS 2015 attracted a relatively small but increasing number of submissions: 16 (qualified) submissions. These submissions reported on up-to-date research findings and application case studies from four countries (China, Australia, USA, and Mexico). A new feature of PAS 2015 was the demo track. After each submission was reviewed by at least three Program Committee members, four full papers, two short papers, and five demo papers were accepted for publication in this volume of conference proceedings (i.e., 25 % acceptance rate for full papers, 12.5 % for short papers, and 31.3 % for demo papers). These 11 papers cover various topics that can be categorized under three main research foci in BPM: process modeling and comparison (four papers), cloud workflow applications (four papers), and process data analysis (three papers).

We would like to thank the Program Committee members for their thorough reviews and discussions of the submitted papers. We express our gratitude to the other conference committees, especially to the general chairs, Jiacun Wang and Qing Wang, and the Steering Committee for their valuable guidance, and to the publicity chairs, Hongyan Zhang, Jihong Liu and Lijie Wen, for their efforts in publishing workshop updates and promoting the workshop in the region. Special thanks to the demo track chairs, Kaijun Ren and Yingbo Liu, for their management and promotion of the new demo track, to the publication chair, Xiao Liu, for his great efforts in preparing the workshop proceedings, and to the organization chair, YuYu Yin, and other staff at Zhejiang University for their attentive preparations for this workshop.

We would also like to take this opportunity to thank the staff at Springer for their efficient work on the publication of the workshop proceedings. Last but not least, we are thankful to the authors of the submissions, the presenters, and all the other workshop participants — the workshop could not be held without their contributions and interest.

November 2015

Jian Cao
He Zhang

Organization

PAS 2015 was organized in Hangzhou, China, by Zhejiang University.

Steering Committee

Jianmin Wang	Tsinghua University, China
Chongyi Yuan	Peking University, China
Liang Zhang	Fudan University, China
Jianwen Su	The University of California, Santa Barbara, USA
Arthur ter Hofstede	Queensland University of Technology, Australia

General Chairs

Jiacun Wang	Monmouth University, USA
Qing Wang	Institute of Software, Chinese Academy of Science, China

Program Chairs

Jian Cao	Shanghai Jiao Tong University, China
He Zhang	Nanjing University, China

Organization Chair

YuYu Yin	Zhejiang University, China

Publicity Chairs

Hongyan Zhang	Beijing Jiaotong University, China
Jihong Liu	Beihang University, China
Lijie Wen	Tsinghua University, China

Demo Track Chairs

Kaijun Ren	National University of Defense Technology, China
Yingbo Liu	Tsinghua University, China

Publication Chair

Xiao Liu	East China Normal University, China

Program Committee

Akhil Kumar	Penn State University, USA
Barbara Weber	University of Innsbruck, Austria
Marco Aiello	University of Groningen, The Netherlands
Massimo Mecella	Sapienza University of Rome, Italy
Rik Eshuis	Eindhoven University of Technology, The Netherlands
Rong Liu	IBM Research, USA
Florian Daniel	University of Trento, Italy
Minseok Song	Ulsan National Institute of Science and Technology, Korea
Matthias Weidlich	Imperial College London, UK
Josep Carmona Vargas	Polytechnic University of Catalonia, Spain
Fabrizio Maria Maggi	University of Tartu, Estonia
Jiacun Wang	Monmouth University, USA
Jian Cao	Shanghai Jiao Tong University, China
Zaiwen Feng	Wuhan University, China
Tao Hu	Hainan University, China
Jianxun Liu	Hunan University of Science and Technology, China
Zongwei Luo	Hong Kong University, Hong Kong, SAR China
Shiyong Lv	Wayne State University, USA
Kaijun Ren	National University of Defense Technology, China
Li Wan	Huazhong University of Science and Technology, China
Jianmin Wang	Tsinghua University, China
Mingzhong Wang	Beijing Institute of Technology, China
Zhongjie Wang	Harbin Institute of Technology, China
Jinhua Xiong	Institute of Computing Technology, CAS, China
Dong Yang	Donghua University, China
Jianwei Yin	Zhejiang University, China
Yang Yu	Sun Yat-Sen University, China
Liang Zhang	Fudan University, China
Yang Zhang	Beijing University of Posts and Telecommunications, China
Xiao Liu	East China Normal University, China
Lizhen Cui	Shangdong University, China
Wanchun Dou	Nanjing University, China
Min Liu	Tongji University, China
Chun Ouyang	Queensland University of Technology, Australia
Raymond Wong	University of New South Wales, Australia
Zhe Shan	University of Cincinnati, USA
Jianwen Su	The University of California, Santa Barbara, USA
Lijie Wen	Tsinghua University, China
Jun Wei	Institute of Software, CAS, China
Jian Yang	Macquarie University, Australia
Kamel Barkaoui	Cedric CNAM Paris, France

Chongyi Yuan   Peking University, China
Luciano Garcia   University of Tartu, Estonia
  Banuelos
Qing Wang   Chinese Academy of Sciences, China
Junchao Xiao   Chinese Academy of Sciences, China
Qi Yu   Rochester Institute of Technology, USA
Xumin Liu   Rochester Institute of Technology, USA
Li Xiong   Emory University, USA
Rong Liu   IBM T.J. Watson Research Center, USA
Jay Shan   University of Cincinnati, USA
Ling Liu   Tsinghua University, China

Contents

Process Modeling and Comparison

Using Classification Method for Querying the Relevant Process Models

Jiaxing Wang, Sibin Gao, Hongjie Peng, Bin Cao, and Jing Fan[✉]

College of Computer Science, Zhejiang University of Technology,
Hangzhou 310000, China
jiaxingwang.zjut@gmail.com, ghzzgd@gmail.com, phjly110@gmail.com,
{bincao,fanjing}@zjut.edu.cn

Abstract. Operations management is important to a company, so more and more business process models are created. At the same time, how to manage such a large amount of process models is becoming a big challenge for companies. Querying the relevant process models is proposed as a business process management technology and it has attracted more and more attention by researchers. The existing methods query the relevant models for a query process model by measuring their similarities. And most of them measure the similarity by focusing on only one kind of feature, such as the structural features or behavioral features, while ignoring other features. In this paper, we consider both structural features and behavioral features to query the relevant process models for a query process model. In order to reach this goal, we use two classification methods named back propagation neural network (BPNN) and support vector machines(SVM) for classifying the candidate models in the repository into two classes: relevant and irrelevant. For the sake of classification, we summarize 7 features to represent the similar or dissimilar parts of two process models. The experiment result shows the precision and efficiency of the classification methods are acceptable.

Keywords: Business process model · BPNN · SVM · Relevant process model · Classification

1 Introduction

Business process models are valuable assets for companies, which results in more and more business process models being created. At the same time, how to effectively manage such a large amount of business process models has become a problem to be solved. A query process model is given to query its relevant models in a business process repository, which has regraded as a technology for managing the business process models. Two process models are relevant, which means they have some similar parts. The majority of the existing similarity methods measure the similarity of two process models by focusing on one of the following three features [1]: (1) text similarity of the labels attached to corresponding nodes (2) structural similarity refers to the graph representation (3) behavioral similarity means their execution sequence of tasks.

© Springer Science+Business Media Singapore 2016
J. Cao et al. (Eds.): PAS 2015, CCIS 602, pp. 3–18, 2016.
DOI: 10.1007/978-981-10-1019-4_1

To find out the relevant process models of a query process model, the first step of previous similarity methods is to compute the similarity scores between it and each solution process model in the repository. Then similarity methods rank the solution process models according to their similarity scores, and regard the top k similar solution process models as relevant process models. That is, there exists a threshold, the similarity scores of top k similar solution process models are greater than it. But there exist some problems in existing methods. The first problem is that they measure the similarity of two process models by only considering one feature, such as structural feature or behavioral feature. As a matter of fact, the similarity of two process models should be measured by multiple features. The second problem is that the weights of each feature is adjusted manually, which is inefficiency and the precision can not be guaranteed. For example, graph edit distance(GED) contains several features such as substitution nodes and substitution edges, and the weights of each feature are adjusted manually.

Unlike previous similarity methods, we use two classification methods named back propagation neural network (BPNN) and support vector machines (SVM) to classify the candidate models in the repository for a query model into two classes: relevant and irrelevant. BPNN is one of the most popular neural network models in practice [2]. BPNN can be easily implemented by using computer programs or circuits [3]. What's more, BPNN has the ability to approximate complicated nonlinear functions. The result of SVM is decided by a small number of support vectors, which are described in Sect. 2, in this way, we can get the key sample data and get rid of the redundant sample data. Besides, the SVM has the ability to deal with the noise, the large data set and the large input space [4]. To our best knowledge, no studies have used machine learning methods for querying the relevant process models. For the sake of classification, we summarize 7 features to represent the similar or dissimilar parts between two process models by considering both structural features and behavioral features. What's more, by using the classification method from machine learning field, the weights of each feature can be adjusted automatically.

Querying the relevant models in the repository for a query model contains five steps. Firstly, the query models and a process model repository with candidate models are input. Secondly, for each query model and each candidate model in the repository, their corresponding 7 features are extracted and each value of feature is computed. Then the feature vector that consists of the 7 feature values is created. The target values of each query model and each candidate model that means whether they are relevant has given through a user study. The third step is to construct the relation vector of each query model and each candidate model. Each relation vector contains their corresponding feature vector and target value. Fourthly, all relation vectors are input to the classification method, where we have manually divided this data into two parts: trained data and tested data. Finally, in the training phase, the classification takes the trained data and generates a classifier. And in the testing phase, the classifier can classify the candidate models for a query model into two classes.

The contribution of this paper is highlighted as follows:

(1) Unlike the previous similarity methods, which need to manually determine the weights of each feature, we use machine learning methods for classification, which can automatically adjust the weights according to different data sets. Hence the classification result will be more suitable for different certain applications.
(2) We can extract an arbitrary feature that represent the similar or dissimilar part of two models for classification. So the features are not limited to the two kinds of features that are mentioned in this paper: structural features and behavioral features, we can consider more kinds of features, such as three kinds: structural, behavioral and semantic features.
(3) We conduct extensively experimental evaluations and the results show that the precision and efficiency of the classification methods are acceptable and BPNN outperforms the Greedy algorithm by only considering structural features in terms of response time.

The rest of the paper is organized as follows. Section 2 presents some preliminaries for our work. Section 3 introduces the 7 features that we summarized and the metrics to calculate the feature values, and our data structure is also introduced. In Sect. 4, the implementation of our idea is described. The experimental evaluations and related work are given in Sects. 5 and 6. Section 7 concludes the paper.

2 Preliminaries

This section introduces some basic concepts that are used throughout the paper. The process models presented in this paper are modeled by Petri-net, so we firstly present the basic Petri net notation. Secondly the basic introduction of support vector machines (SVM) and back propagation neural network (BPNN) are presented. Finally, we formally describe the problem that is to be solved in this paper.

2.1 Petri-net Based Process Modeling

A Petri net consists of two kinds of basic nodes: place and transition, where transition represents the activity or task and place means the state or condition of its neighbouring transitions. The directed arcs connect place with transition or transition with place, and the same type of two nodes are not allowed to connect. The elements of Petri net can be seen in Fig. 1. A place contains a number of tokens that represent resource. For a transition A, the places that flow into it through arcs are called A's input places, and these places that flow out it are called A's output places. Transition A can be fired if its input places contain enough tokens, which results in the tokens are transferred from A's input places to A's output places.

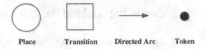

Place   Transition   Directed Arc   Token

Fig. 1. The graphical representation of Petri-net elements

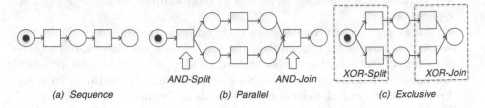

AND-Split

AND-Join

XOR-Split

XOR-Join

(a) Sequence   (b) Parallel   (c) Exclusive

Fig. 2. Basic control flow constructs modelled by Petri-net

A process model usually describes which tasks need to be executed and in what order. The transition in the process model is the indivisible unit that must be completely executed. The control constructs of a Petri net based process model are made of the combination of connected transitions and places, which reflect the routing modes: sequence, parallel and exclusive [6], as shown in Fig. 2.

Sequence: As Fig. 2(a) shows, it is a causal relation between nodes which is represented by adding a place node between two transitions.

Parallel: AND-Split can execute several tasks at the same time while AND-Join is fired after its all preceding tasks have executed, as shows in Fig. 2(b).

Exclusive: As Fig. 2(c) shows, the process model contains two paths that start at the same place and immediately diverge, and the XOR-Split chooses one of the two paths to go.

2.2   Classification Method

Classification is widely used in machine learning, for example, an E-mail system will estimate whether an E-mail is a spam by classification method. In order to classify the candidate process models into two classes, we use two classification methods called support vector machine(SVM) and back propagation neural network(BPNN).

Support Vector Machine. The interesting process models to the users are only a very small portion in the large business process repository, so the problem of querying the relevant process models can be regarded as a strict two-class classification problem. The support vector machine classifier uses a binary classifier algorithm that looks for an optimal hyperplane as a decision function in a high-dimensional space [7]. As Fig. 3 shows, we suppose the circles are the relevant models, the squares are the irrelevant models, and the decision boundary (the dark lines between two dashed lines) will classify them into two classes.

However, there exist many decision boundaries that can separate two classes, and the goal of SVM is to find the decision boundary with the maximal margin, in Fig. 3, the decision boundary is the bold solid line. The margin means the distance between the two dash lines, and SVM selects the decision boundary that makes the distances from two dash lines to the decision boundary are identical. The data points with the black boxes are on the boundary of margin called support vectors, which are also two classes and connected by two dash lines respectively [4].

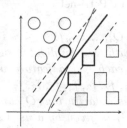
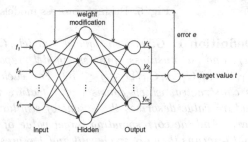

Fig. 3. The decision boundary of SVM in a two-dimensional feature space   **Fig. 4.** A schematic of a back-propagation neural network

Back Propagation Neural Network. The back propagation neural network (BPNN) is a multilayered feedforward network based on error back propagation, which is one of the most widely used neural network models [2]. The input of BPNN is a vector $input = \{f_1, f_2, ..., f_n, t\}$. Thereinto, $f_1, f_2, ..., f_n$ means the features that can represent an object, and t means the target value or a correct classification. For example, a song can be represented by 4 features: $(title, singer, mood, duration)$, and the target value is 1 if its style is Jazz while 0 means other style. In order to classify the objects into certain types, each feature $f_1, f_2, ..., f_n$ is weighted by $x_1, x_2, ..., x_n$ respectively. BPNN minimizes the squares sum of error by repeatedly adjusting the weight vector $\{x_1, x_2, ..., x_n\}$.

As Fig. 4 shows, the BPNN generally contains three layers: input layer, hidden layer and output layer. The BPNN learning process can be divided into two phases [8]. Firstly, the vector $input = \{f_1, f_2, ..., f_n, t\}$ is input, $f_1, f_2, ..., f_n$ is processed from input layer to hidden layer in order to reach the target value t, then the result is transformed to the output layer, accordingly the error e is produced. Secondly, if e is not enough small to accept, the error signal is back to all layers and their weights are updated. Otherwise, the weight vector is produced.

2.3   Problem Description

For a query model and a business process model repository, the goal of this paper is to classify the candidate models in the repository for the query model into two classes: relevant and irrelevant. Next We formally define the problem as follows.

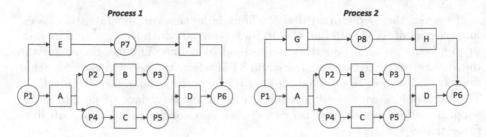

Fig. 5. Business process models modeled by Petri-net

Definition 1. *Given m query model $Q_1, Q_2, ..., Q_i, ..., Q_m$ ($1 \leq i \leq m$) and a business process model repository with n candidate models $C_1, C_2, ..., C_j, ..., C_n$ ($1 \leq j \leq n$). For each Q_i and each C_j, a feature vector is constructed, which consists of 7 feature values: $f_1, f_2, f_3, f_4, f_5, f_6, f_7$. Each feature value describes the extent of similarity or dissimilarity that Q_i and C_j have. And the corresponding target value of Q_i and C_j is given by a user study that 0 means they are irrelevant and 1 represents relevant. $m \times n$ feature vectors and their corresponding target values are input to the classification method. The input data is divided into two parts: trained data and tested data. The trained data is used to adjust the weights of each feature to reach the target value. The tested data is to be tested whether the classification result is right.*

3   Process Features

We use the term of feature that from the machine learning field to represent the feature of process model, such as structural features or behavioral features. For example, in machine learning, we can use 3 features to represent a car: $Car = \{Price, Speed, surface\}$. In this paper, the features of process model are divided into two classes: structural features and behavioral features. Structural features contain 4 features: substitution nodes, deletion/insertion nodes, substitution edges, deletion/insertion edges. And behavioral features consist of causal relations, conflict relations and concurrency relations. In this section, we firstly describe the 7 features in detail, then the metrics about how to calculate their corresponding feature values are proposed. Finally, based on the feature values, the feature vector is proposed. And our data structure is created by using the feature vector.

3.1   Feature Description

Structural Features. A process model is usually represented as a graph, where nodes stand for tasks and edges encode causal dependencies between adjacent nodes. Therefore, the structure is an important feature for a process model. If two process models are relevant, then they may be partly similar in structure. We referenced the graph edit distance [9] and extracted 4 specific features on structure of two process models: substitution nodes, deletion/insertion nodes, substitution edges, deletion/insertion edges.

In order to introduce these notions in detail, in this section we take Fig. 5 as an example to illustrate. There are two Petri-net process models named *"Process 1"* and *"Process 2"* in Fig. 5. Obviously, the difference of *"Process 1"* and *"Process 2"* is the top path. The top path in *"Process 1"* is *"P1 → E → P7 → F → P6"*, while in *"Process 2"* the top path is *"P1 → G → P8 → H → P6"*. And the rest of them is the same. Next we detailed describe each feature.

Feature 1: Substitution Nodes. Two nodes are substitution nodes, which means a node from one process model is substituted for a node from the other process model. This also means there exists a corresponding relation between the two nodes. The relation between two substitution nodes can be measured by text similarity [13] or semantic similarity [14]. Whether two nodes are substitution nodes or not depends on different applications. In some applications, two nodes are substitution nodes if they have the maximum similarity. In some other applications, two nodes are substitution nodes if their labels are identical.

In Fig. 5, we suppose two nodes are substitution nodes if their labels are identical. So the substitution nodes of *"Process 1"* and *"Process 2"* are $\{A, B, C, D, P1, P2, P3, P4, P5, P6\}$, where $\{A, B, C, D\}$ are the substitution transitions (tasks) and $\{P1, P2, P3, P4, P5, P6\}$ are substitution places.

Feature 2: Deletion/Insertion Nodes. After all the substitution nodes have been identified in two process models, the rest nodes are deletion or insertion nodes. As Fig. 5 shows, the deletion/insertion nodes are $\{E, P7, F, G, P8, H\}$.

Feature 3: Substitution Edges. An edge has two endpoints: a left node and a right node. Two edges are substitution edges, that is, an edge from one process model is substituted for an edge from other process model. In other words, two edges are substitution edges if their corresponding left nodes are substitution nodes and right nodes are also substitution nodes. As Fig. 5 shows, the substitution edges are $\{P1 → A, A → P2, P2 → B, B → P3, P3 → D, D → P6, A → P4, P4 → C, C → P5, P5 → D\}$.

Feature 4: Deletion/Insertion Edges. Two edges from two models are deletion or insertion edges if their corresponding left nodes or right nodes, or both left and right nodes are not substitution nodes. As Fig. 5 shows, the deletion/insertion edges are $\{P1 → E, E → P7, P7 → F, F → P6, P1 → G, G → P8, P8 → H, H → P6\}$.

Behavioral Features. For a process model, besides its static topology, it also contains the execution sequence of task nodes, which is called behavior. The behavior is also an essential feature, which reflects the execution relations between two task nodes of a process model: causal relation, conflict relation, and concurrency relation. Two process models are relevant, they possibly share some common behavioral features. Next we introduce each behavior feature in detail.

Feature 5: Causal Relation. Two task nodes from a process model are in causal relation, which means their execution order is sequential, that is, one task node can be executed only after another task node has been executed. As Fig. 5 shows, the causal relations of "*Process 1*" are $\{A \rightarrow B, A \rightarrow D, A \rightarrow C, B \rightarrow D, C \rightarrow D, E \rightarrow F\}$, and $\{A \rightarrow B, A \rightarrow D, A \rightarrow C, B \rightarrow D, C \rightarrow D, G \rightarrow H\}$ are the causal relations of "*Process 2*", where "$\rightarrow$" means the sequential relation.

Feature 6: Conflict Relation. If two task nodes in a process model are in conflict relation, then only one of the two task nodes is chosen to be executed. In Petri-net, two tasks are in conflict relation if they start at the same place and immediately diverge. As Fig. 5 shows, the conflict relations in "*Process 1*" are $\{A\#E, A\#F, B\#E, B\#F, C\#E, C\#F, D\#E, D\#F\}$, and $\{A\#G, A\#H, B\#G, B\#H, C\#G, C\#H, D\#G, D\#H\}$ are the conflict relations of "*Process 2*", where we use "$\#$" to describe the conflict relation.

Feature 7: Concurrency Relation. Two task nodes from a process model are in concurrency relation if they are neither in causal relation nor in conflict relation, that is, they can be executed at the same time. As Fig. 5 shows, the concurrency relation of "*Process 1*" is $\{B == C\}$, and $\{B == C\}$ is also the concurrency relation of "*Process 2*", where "$==$" means the concurrency relation.

3.2   Feature Metrics

In this paper, we extracted 7 features between two process models from two classes of features: structural features and behavioral features, and the value of a certain feature can partly describe the degree of similarity of two process models. The metrics about how to calculate the feature values are defined as follows:

Definition 2. *Given two process models Q and C, StructuralFeature(Q, C) represents the 4 structural features, that is, substitution nodes or deletion/insertion nodes or substitution edges or deletion/insertion edges, and the corresponding feature value is denoted as: $StructuralFeatureValue$ $(StructuralFeature(Q, C)) = \frac{|StructuralFeature(Q,C)|}{|NodeOrEdge(Q) \cup NodeOrEdge(C)|}, 0 \leq Structural$ $FeatureValue \leq 1$.*

In *Definition 2*, | *StructuralFeature(Q, C)* | is the size of *StructuralFeature(Q, C)*. For example, *StructuralFeature(Q, C)* represents the substitution nodes of two process models, correspondingly, | *StructuralFeature(Q, C)* | means the size of their substitution nodes. *NodeOrEdge(Q)* represents the set of nodes or edges in Q. If *StructuralFeature(Q, C)* refers to the feature of nodes such as substitution nodes or deletion/insertion nodes, then *NodeOrEdge(Q)* ∪ *NodeOrEdge(C)* is the union of Q's node set and C's node set, the size of which is represented by | *NodeOrEdge(Q)* ∪ *NodeOrEdge(C)* |. Similarly, if substitution edges or deletion/insertion edges is referred by *NodeOrEdge(Q)*, then $NodeOrEdge(Q) \cup NodeOrEdge(C)$ is the union of Q's edge set and C's edge set.

For example, in Fig. 5, the value of **Feature 1** of "*Process 1*" and "*Process 2*" is $StructuralFeatureValue(Substitution\ Nodes) = \frac{10}{16}$, the value of **Feature 2** is $StructuralFeatureValue(Deletion\ and\ Insertion\ Nodes) = \frac{6}{16}$, the value of **Feature 3** is $StructuralFeatureValue(Substitution\ Edges) = \frac{10}{18}$, and the value of **Feature 4** is $StructuralFeatureValue(Deletion\ and\ Insertion\ Edges) = \frac{8}{18}$.

Definition 3. *Let Q and C be two process models, BehavioralFeature(Q) and BehavioralFeature(C) be their corresponding behavioral features respectively. The value of the corresponding behavioral feature between Q and C is defined as follows: $BehavioralFeatureValue(BehavioralFeature(Q), BehavioralFeature(C)) = \frac{|BehavioralFeature(Q) \cap BehavioralFeature(C)|}{|BehavioralFeature(Q) \cup BehavioralFeature(C)|}$, $0 \le BehavioralFeatureValue \le 1$.*

For example, in Fig. 5, the value of **Feature 5** of "*Process 1*" and "*Process 2*" is $BehavioralFeatureValue(Causal\ Relation) = \frac{5}{7}$, the value of **Feature 6** is $BehavioralFeatureValue(Conflict\ Relation) = 0$, and the value of **Feature 7** is $BehavioralFeatureValue(Concurrency\ Relation) = 1$.

3.3 Feature Vector

For two process models, in order to measure whether they are relevant, we extracted 7 features that represent the similar or dissimilar parts of them, the values of which are ranged from 0 to 1. The bigger the feature value is, the more similar or dissimilar the corresponding part of two process model is. We create a feature vector to store the 7 features. A feature vector is denoted as $FV = (Feature\ 1, Feature\ 2, Feature\ 3, Feature\ 4, Feature\ 5, Feature\ 6, Feature\ 7)$.

For example, in Fig. 5, the feature vector of "*Process 1*" and "*Process 2*" is $FV(Process\ 1,\ Process\ 2) = (0.625,\ 0.375,\ 0.556,\ 0.444,\ 0.714,\ 0,\ 1)$.

Based on the feature vector, we create a relation vector as our data structure that is called *Relation Vector*, and a relation vector is denoted as $RV = (FV,\ target)$. *target* means whether two process models are relevant or not is given by a user study, where *target = 0* represents they are irrelevant and *target = 1* means they are relevant.

4 Implementation

Based on the problem description, we use two classification methods named BPNN and SVM for querying the relevant process models. In this section, we firstly describe the details of feature extraction, then the steps of querying the relevant process models by using classification method are represented.

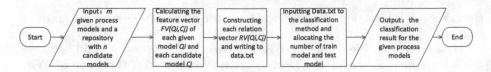

Fig. 6. Five steps of querying the relevant process models by using classification method

4.1   Feature Extraction

In this paper, we have summarized 7 features to represent the similar or dissimilar parts between two process models by considering structural features and behavioral features. Next we describe the implementation of feature extraction for two process models, which consists of 3 steps.

Step 1: The substitution nodes of two process models are determined, that is, a mapping of nodes is constructed. We apply the greedy algorithm [5] to this implementation. Firstly, the similarity scores of all possible pairs of nodes are calculated, where one node from one process model and another node from the other process model. And the type of pair of node must be the same, for example, in Petri net, place must map with place and task map with task. Then the greedy algorithm selects an pair of nodes that their similarity score can most increase the total similarity in each iteration. Finally, no similarity score of pair of node can increase the total similarity, then the substitution nodes are determined.

Step 2: After the substitution nodes are found out, the rest nodes in two process models are deletion or insertion nodes. Correspondingly, each pair of edge that one edge from one process model and another edge from another process model can be examined whether they are substitution edges. In this pair of edge, if the pair of left endpoint nodes and the pair of right endpoint nodes are all substitution nodes, then they are substitution edges. Similarly, the rest edges in two process models are deletion or insertion edges.

Step 3: We use a jar package named jbpt.jar [10] that can derive behavioral profiles of a Petri net, which has different methods for the computation. By using jbpt.jar, we can obtain the causal relation, conflict relation and concurrency relation of two process models.

4.2   Querying the Relevant Process Models

The implementation of using the classification method for querying the relevant process models contains five steps, as shown in Fig. 6.

Step 1: m query models and a repository with n candidate models are input.

Step 2: For each query model Q_i and each candidate model C_j, $1 \leq i \leq m$ and $1 \leq j \leq n$, their corresponding feature vector $FV(Q_i, C_j)$ that consists of 7 feature values is calculated.

Step 3: Each *target* value of Q_i and C_j that means whether they are relevant or not has given by a user study in the data set, so the relation vector $RV(Q_i, C_j)$

consists of feature vector and *target* value is constructed. Then each $RV(Q_i, C_j)$ is written to a file named data.txt, that is, at last there are $m \times n$ records in data.txt.

Step 4: The data.txt is input to the classification method, where we have divided the input data into two parts: trained data and tested data. For example, there are 10 query models and a repository with 100 candidate models, that is, there exists 1000 records in data.txt. We can allocate 5 query models that refer to 500 records in data.txt to be trained, and the rest 5 query models and their corresponding 500 records are to be tested.

Step 5: The classification method output the classification result, that is, for each test query model, the candidate models in the repository are classified into two classes for it: relevant and irrelevant.

5   Experiments

In this section, two kinds of experimental evaluations are provided. We firstly compare the classification precision of BPNN with SVM by considering structural features and behavioral features. The comparison between BPNN and SVM is in two ways: one is fixing the number of trained query models and varying their combination, and the other is varying the number of trained query models. Secondly, we compare the query quality and response time of BPNN with the Greedy algorithm by only considering structural features. All experiments in this section are based on the real data set from IBM [11], which contains more than 3000 business process models. And we manually chose several query models for querying the repository in terms of the control constructs and modified a set of relevant models for them. All experiments were evaluated on a machine with Intel(R) Xeon(R) CPU E5-2637 v2, 3.5 GHz processor and 8 GB RAM, running JDK 1.7 and Windows 7.

5.1   BPNN Vs SVM

This section compares the classification quality of BPNN with SVM by considering both structural features and behavioral features. At first, we investigate the impact of the combination of trained query models by fixing their number. Then we vary the number of trained query models. To compute the precision, we

Table 1. Static data of query models

Query model id	1	2	3	4	5	6	7	8	9	10
place number	13	16	37	39	59	38	76	85	39	43
task number	9	10	41	43	31	31	54	77	31	35
edge number	24	30	82	86	117	74	161	183	82	92
control structure	Seq	Seq	OR	OR	AND	AND	Loop	Loop	AND+OR	AND+OR

Table 2. The combination of trained models

Combination id	1	2	3	4	5	6	7
Figure 7	13579	2468,10	12345	6789,10	34567	1289,10	45678
Figure 8	13579,10	24689,10	123456	56789,10	345678	12489,10	456789
Figure 9	456789,10	2346789	134579,10	1234567	1245689	246789,10	1235679

choose 10 query models from the data set with different complexities and control structures. As Table 1 shows, the selected query models are numbered from 1 to 10, where each two models contains a same control structure. The selected query models numbered 1 and 2 contain the sequence structure(Seq), 3 and 4 have the conflict structure(OR), the concurrency structure(AND) is contained by 5 and 6, 7 and 8 contain a loop structure(Loop), the control structure of 9 and 10 consist of conflict and concurrency(AND + OR).

We manually modify each of them to 9 variants by considering structural and behavioral features. In this way, together with 10 query models, we build a business process repository with 100 models. Therefore, each query model has 10 relevant process models in the repository(9 variants and itself).

Figures 7, 8 and 9 show the precision scores of classification by using BPNN and SVM, where the number of trained query models is 5, 6 and 7 respectively, that is, 500 records that calculated by 5 query models and 100 candidate models, 600 records that calculated by 6 query models and 100 candidate models and 700 records that calculated by 7 query models and 100 candidate models are trained respectively. And their corresponding combination is shown in Table 2. On the whole, BPNN and SVM can achieve a good precision and BPNN outperforms the SVM. As shown in Fig. 7, five query models are trained. The first and second combination achieve the highest precision, by referencing Table 2 we can know their corresponding combinations of trained query models are *1, 3, 5, 7, 9* and *2, 4, 6, 8, 10*. From the structural aspect, the two combinations with the highest precision contain all control structures while other combinations don't have all the control structures, as shown in Tables 1 and 2. Therefore, the classification methods can learn from the trained query models well and give a good classification result.

In Fig. 8, six query models are trained. The number of trained models is bigger than Fig. 7 but the precision scores of some combinations are lower. The combination id with the lowest precision score is 4, next are 5 and 7. From the structural aspect, as shown in Table 1, the 4th combination trains the query models numbered *5, 6, 7, 8, 9, 10*, 5th and 7th combinations train *3, 4, 5, 6, 7, 8* and *4, 5, 6, 7, 8, 9* models respectively. We can see that the latter two combinations contain no sequence structure and the former contains no sequence and conflict structure.

The result of training seven query models are shown in Fig. 9, where the precision scores of all combinations are high. After all, the training set should contain all the features as far as possible. The bigger the number of trained

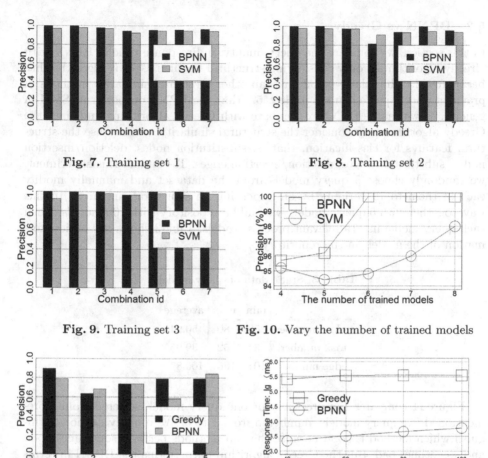

Fig. 7. Training set 1   **Fig. 8.** Training set 2

Fig. 9. Training set 3   **Fig. 10.** Vary the number of trained models

Fig. 11. Precision   **Fig. 12.** Vary the repository size

models is, the more possible all features are selected, correspondingly the precision will be higher.

Figure 10 compares the classification quality of BPNN with SVM by varying the number of trained query models. We can see that the two classification methods can get good precision and BPNN outperforms SVM, and with the increment of trained query models, the precisions are overall getting higher. But two combinations in SVM perform poor, which are *1, 2, 8, 9, 10* and *1, 2, 4, 8, 9, 10*. The reason is that SVM is more sensitive to features than BPNN, once the features in SVM become the support vectors, then these features contribute more than others.

5.2 BPNN vs Greedy

In this section, we compare the query quality and response time of BPNN with Greedy algorithm by only considering structural features. Greedy algorithm has been proved to be relatively better than other structural similarity methods in previous work [5]. We can see in Sect. 5.1 that the BPNN outperforms SVM in general, so we choose BPNN to compare with Greedy algorithm. And since the Greedy algorithm only consider the structural similarity, we only use the structural features for classification, that is, substitution nodes, deletion/insertion nodes, substitution edges, deletion/insertion edges. To conduct this experiment, we randomly choose 5 query models from the data set and manually modify each of them to 19 variants by only focusing on structural similarity. In this way, together with 5 query models, we build a repository with 100 models. Then each query model has 19 relevant process models. Table 3 shows the basic information of the models in the repository.

Table 3. Static data of query models

	min	max	average
place number	50	80	50.94
task number	34	52	39.95
edge number	101	166	106.87

Figure 11 compares the precision for one query model, selected from aforementioned five query models, numbered from 1 to 5. We roughly cannot distinguish which method is good because the two methods perform sometimes good and sometimes bad. But the Greedy algorithm outperforms the BPNN in terms of average precision. However, BPNN has an opportunity for improvement in terms of precision, that is increasing the number of trained models. The more query models are trained, the weights of each feature will be adjusted more accurate.

In Fig. 12, the repository size is varied from 40 to 100, we observe the changes of BPNN and Greedy. Generally, the query response time for both methods grow as the repository size increases and BPNN is more quicker than Greedy.

In summary, using BPNN and SVM for querying the relevant models by 7 features in terms of structure and behavior can achieve a acceptable precision. The more features the training set contains, the higher the precision is. BPNN can achieve almost the same precision as the Greedy algorithm, and it has a large space for improvement by increasing the number of trainded query models. What is more important, BPNN outperforms the Greedy algorithm in terms of the query response time for all cases. Thus we can roughly consider our feature based classification methods can satisfy the users' real needs.

6   Related Work

The existing methods query the relevant models by measuring the similarity, and regard the top k similar models as relevant models. Most of them measure the similarity by focusing on one of the following three features: text similarity, structural similarity and behavioral similarity. J. Wang et al. [12] provided a good overview on the business process model similarity search from the above mentioned three aspects.

Weidlich et al. [13] measured the similarity by considering the text similarity of two process models. Firstly, a mapping is built to identify which node in one model corresponds to which node in another model. Then the rate of mapped nodes is to measure two models' similarity. R. Dijkman et al. [5] compared four algorithms in terms of structural similarity by using graph edit distance. The Greedy algorithm performs the best from computational complexity and the retrieval quality. Cao et al. [15] used the Hungarian algorithm for mapping the elements, which can save a lot of time for searching the best combination of elements mapping. Zha et al. [16] proposed a similarity measurement based on the transition adjacency relation set to compare the behavior similarity between business process models.

For other related work, Yan et al. [17] selected small characteristic model fragments, called features, which were used to estimate model similarities and classified them as relevant, irrelevant or potentially relevant to a query model. Potentially relevant process models are compared by using a full graph comparison technique. And the features refer to the structural aspects. Jin et al. [18] proposed a method to compute the semantic features of business process models and used indexes to support for the query processing. Cao et al. [19] used the maximal common subgraph approach instead of graph edit distance for comparing the similarity of process models.

7   Conclusion

The start point of this paper is the features of two process models, which mean the similar or dissimilar parts of them. In this paper, we have summarized 7 features by considering structural features and behavioral features. Then based on these features, we use two classification methods named BPNN and SVM for querying the relevant models for a query process model. The experimental evaluation shows that BPNN has a higher precision than SVM, and with the increment of trained query models, the precisions of the two classification methods are getting higher. What's more, BPNN outperform the Greedy algorithm by almost 1.7 orders of magnitude in terms of the query time. Therefore, the precision and efficiency of our feature based classification methods can meet the need of people. The shortage of this paper is that we can only classify the candidate models into two classes: relevant and irrelevant, but we cannot rank the relevant models according to their relevance, which is also our future work.

References

1. Dijkman, R.M., Dumas, M., Dongen, B.V., Käärik, R., Mendling, J.: Similarity of business process models: metrics and evaluation. Inf. Syst. **36**(2), 498–516 (2011)
2. Fausett, L.: Fundamentals of Neural Networks: Architectures, Algorithms, and Applications. Prentice-Hall Inc., Upper Saddle River (1994)
3. Wong, W.E., Qi, Y.: BP neural network-based effective fault localization. Int. J. Softw. Eng. Knowl. Eng. **19**(04), 573–597 (2009)
4. Zavaljevski, N., Stevens, F.J., Reifman, J.: Support vector machines with selective kernel scaling for protein classification and identification of key amino acid positions. Bioinformatics **18**, 689–696 (2002)
5. Dijkman, R., Dumas, M., García-Bañuelos, L.: Graph matching algorithms for business process model similarity search. In: Dayal, U., Eder, J., Koehler, J., Reijers, H.A. (eds.) BPM 2009. LNCS, vol. 5701, pp. 48–63. Springer, Heidelberg (2009)
6. van Der Aalst, W.M., Ter Hofstede, A.H., Kiepuszewski, B., Barros, A.P.: Workflow patterns. Distrib. Parallel Databases **14**(1), 5–51 (2003)
7. Joachims, T.: Making Large Scale SVM Learning Practical. Universität Dortmund, Dortmund (1999)
8. Tam, K.Y., Kiang, M.Y.: Managerial applications of neural networks: the case of bank failure predictions. Manag. Sci. **38**(7), 926–947 (1992)
9. Bunke, H.: On a relation between graph edit distance and maximum common subgraph. Pattern Recogn. Lett. **18**(8), 689–694 (1997)
10. https://code.google.com/p/jbpt/wiki/BasicHowTo
11. Fahland, D., Favre, C., Jobstmann, B., Koehler, J., Lohmann, N., Völzer, H., Wolf, K.: Instantaneous soundness checking of industrial business process models. In: Dayal, U., Eder, J., Koehler, J., Reijers, H.A. (eds.) BPM 2009. LNCS, vol. 5701, pp. 278–293. Springer, Heidelberg (2009)
12. Wang, J.M., Tao, J., Wong, R.K., Wen, L.J.: Querying business process model repositiories. World Wide Web **17**, 427–454 (2014)
13. Weidlich, M., Dijkman, R., Mendling, J.: The ICoP framework: identification of correspondences between process models. In: Pernici, B. (ed.) CAiSE 2010. LNCS, vol. 6051, pp. 483–498. Springer, Heidelberg (2010)
14. Ehrig, M., Koschmider, A., Oberweis, A.: Measuring similarity between semantic business process models. In: Proceedings of the Fourth Asia-Pacific Conference on Comceptual Modelling-vol. 67, pp. 71–80. Australian Computer Society Inc. (2007)
15. Cao, B., Wang, J., Fan, J., et al.: Mapping elements with the hungarian algorithm: an efficient method for querying business process models. In: 2005 IEEE International Conference on Web Services (ICWS), pp. 129–136. IEEE (2015)
16. Zha, H., Wang, J., Wen, L.J., Wang, C., Sun, J.: A workflow net similarity measure based on transition adjacency relations. Comput. Ind. **61**, 463–471 (2010)
17. Yan, Z., Dijkman, R., Grefen, P.: Fast business process similarity search with feature-based similarity estimation. In: Meersman, R., Dillon, T.S., Herrero, P. (eds.) OTM 2010. LNCS, vol. 6426, pp. 60–77. Springer, Heidelberg (2010)
18. Jin, T., Wang, J., Wen, L.: Querying business process models based on semantics. In: Yu, J.X., Kim, M.H., Unland, R. (eds.) DASFAA 2011, Part II. LNCS, vol. 6588, pp. 164–178. Springer, Heidelberg (2011)
19. Cao, B., Yin, J., Li, Y., Deng, S.: A maximal common subgraph based method for process retrieval. In: 2013 IEEE 20th International Conference on Web Services (ICWS), pp. 316–323. IEEE (2013)

An Approach Towards User Interface Derivation from Business Process Model

Lei Han[⊠], Weiliang Zhao, and Jian Yang

Macquarie University, Sydney, NSW 2109, Australia
lei.han1@students.mq.edu.au, {weiliang.zhao,jian.yang}@mq.edu.au

Abstract. This paper proposes an approach for user interface (UI) generation and updating. A role-enriched business process model is developed with detailed description for tasks and associated data. The model is specified in an extended BPMN. A set of control flow patterns and data flow patterns are identified based on the proposed model for UI derivation. A comprehensive set of constraints and recommendations are specified for supporting the UI generation and updating. This early work will lay a foundation towards an effective tool for supporting UIs development and maintenance.

Keywords: User interface · Role · Data relationship · Business process

1 Introduction

Business Process Management Systems (BPMSs), which separate the management of business processes from application softwares of enterprises, facilitate enterprises to analyze, design, execute, and monitor business processes [1,2]. User interfaces of business processes serve an interaction bridge between end users and business processes (BPs). Through the UIs, end users can retrieve data stored in a system and provide input data to the system. The UI development is usually labour-intensive and error-prone. Normally BPs and UIs are tightly coupled with each other. The realization and maintenance of UIs are often costly and effort-consuming, which impedes the quick adaptations of business process realization [3].

It is desirable to have mechanics that the UI logic can be derived from business process models. In order to support UI derivation, the current business process models need to be enriched. We believe that a UI should exhibit the following features: (1) each participating role has a UI; (2) operation flow is clearly specified; (3) data items specified in a UI are consistent with operation orders; (4) each data item of a UI is specified with access type, i.e., read, write, update, with options: default, compulsory or optional.

In order to derive the complex UI logic, we propose a UI derivation approach with the following features:

- A role-enriched business process model is proposed with operations, data items, and associated operation orders.

© Springer Science+Business Media Singapore 2016
J. Cao et al. (Eds.): PAS 2015, CCIS 602, pp. 19–28, 2016.
DOI: 10.1007/978-981-10-1019-4_2

- A rich set of control flow patterns and data flow patterns are identified to describe operation flows of tasks within a BP and operation orders of data within a task respectively.
- A set of constraints and recommendations are proposed as UI derivation rules for analysing, generating, and updating user interfaces.

The remainder of this paper is organized as follows. Section 2 provides an overview of the proposed approach for UI derivation. Section 3 proposes a role-enriched business process model. Section 4 describes the process of business process abstraction and aggregation. Section 5 specifies data relationships and their derivation. Section 6 presents a UI model and discusses UI derivation rules. Section 7 reviews some existing works. Section 8 gives a conclusion.

2   UI Derivation Framework

UIs are used to realize the interaction between end users and a BP. The proposed approach derives a UI model for each participating user role. This UI model comprises a set of UIs and operation flows between them. Each UI contains a set of data items and associated operation orders. Each data item needs to be specified with an access type and its option. The access type indicates if a data item is to be read or edited, while the option represents if the operation of this data item is compulsory or optional. The proposed derivation framework is as follows (see Fig. 1):

- In order to derive UIs, the BP model should have features as: (1) Each task of a BP needs to be specified with user roles to be involved in this task; (2) each task of the BP is specified with operated data and their operation orders, which will be used to derive the operation orders of data items in a UI and execution flow of multiple UIs. A rich set of control flow patterns, which represent the operation flows of tasks within a BP, is specified. The BPs are abstracted and aggregated for each user role based on these patterns. Meanwhile, the relationships of data operated by tasks are derived.
- At the first step, the business process is abstracted and aggregated for each user role. The tasks used only by other roles are hidden according to the identified control flow patterns of the BP.
- At the second step, the logic of task flows and data flows within tasks is derived. Ten elementary types of data relationships are identified based on control flow patterns. These elementary types are used as the foundation to represent complex data relationships.
- Finally, UI derivation is carried out based on the obtained data relationships at the second step. UI container, UI model, and UI constraints and recommendation rules are specified to support the user interface generation and updating.

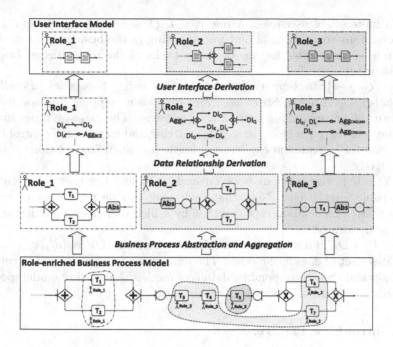

Fig. 1. Derivation framework from role-enriched business process model to UI model

3   Role-Enriched Business Process Model

This section proposes a role-enriched business process model, which is the starting point of this derivation framework. In order to derive the complex UI logic for each individual user role, the process model is specified with the relations between user roles and tasks and detailed description of data flow relations within each task. A set of control flow patterns are identified as the basis of the UI derivation. We will firstly provide the formal definition of role-enriched business process model. Then, we will specify a set of control flow patterns.

3.1   Formal Definitions

A role-enriched business process model consists of the task flow relations with details of attribute operations in individual tasks. The definition for the role-enriched business process model are as follows:

Definition 1: Role-enriched Business Process Model. A role-enriched business process model is denoted as a tuple $rm = (N_t, R, type_G^t, CF_{ex}^t, CF_{im}^t, \rho, DF, refine)$, where:

- $N_t = \{e_s^t, e_e^t\} \cup G_t \cup T$ is a finite set of flow objects. e_s^t, e_e^t, and G_t indicate start event, end event, and a finite set of gateways, respectively. T is a finite set of tasks of the role-enriched business process model. A task $t \in T$ indicates a logic unit of work.

- R is a finite set of user roles. A role $r_i \in L$ $(1 \leqslant i \leqslant N)$ represents a group of users or an organizational unit participating in the business process. Each user role is responsible for one or many tasks of the role-enriched business process model.
- $type_G^t$: G_t $\rightarrow$ {$Strict$-$order$ $Sequential$, $Free$-$order$ $Sequential$, $Parallel$-A, $Parallel$-B, $Conditional$, $Strict$-$loop$, $Free$-$loop$} is a mapping function used to indicate the corresponding type of each gateway. The gateway types include entering and exiting a block, as well as splitting and merging the control flows.
- CF_{ex}^t and CF_{im}^t represent explicit and implicit control flow relations of tasks, respectively.
- $\rho = T \times R$ is the relations between each task and the user roles who are responsible for the task. Note that each task can be performed by one or many roles. This relation captures that by which user roles a particular task is to be performed.
- $refine$: $T \rightarrow DF$ is a refinement function on the tasks. $DF = \{df_1, df_2, ..., df_n\}$ indicates a set of data flow models. The refinement function is used to connect the tasks and their corresponding data flow models. A data flow model specifies the orders of attribute operations within a task.

3.2   Control Flow Patterns

Figure 2 demonstrates the control flow patterns of tasks. There are two kinds of sequential patterns: *Strict-order Sequential* and *Free-order Sequential*. The *Strict-order Sequential* specifies that tasks are executed strictly in sequential order, and the execution order cannot be freely changed; the *Free-order Sequential* indicates that tasks are executed one after another, however the execution can be in any order. The parallel patterns includes two types: *Parallel-A* and *Parallel-B*. The *Parallel-A* represents all the branches are executed in parallel, and must be all completed; all the branches in the *Parallel-B* are executed in parallel, however, only one of the branches is needed to be complete. One of the branches in *Conditional* is executed based on its real-time situation. For loop patterns, there exist *Strict-loop* and *Free-loop*. All the tasks in the *Strict-loop* are executed in strict-order sequential during each iteration. All the tasks in the *Free-loop* are executed in free-order sequential during each iteration.

4   Business Process Abstraction and Aggregation

This section describes the business process abstraction and aggregation, which is the first step of our UI derivation process. The abstraction and aggregation of business process refer to hiding the irrelevant tasks and aggregate them into abstract nodes for a user role. Based on the abstracted and aggregated business process, the role-specific UIs will be generated for each individual user role.

In the abstraction and aggregation, the tasks to be executed by the user role must be reserved, while the tasks irrelevant to this role will be aggregated to abstract nodes. Fourteen patterns are identified, they are based on our previously introduced control flow patterns. Due to the space limit of this paper, we

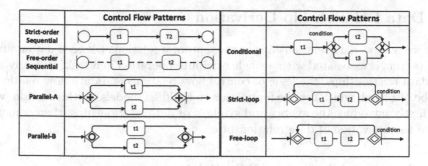

Fig. 2. Control flow patterns of role-enriched business process model

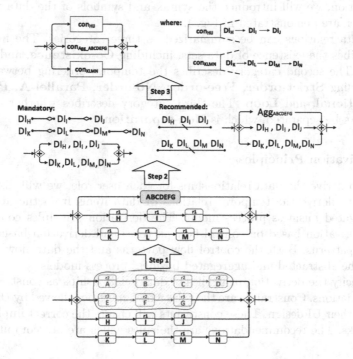

Fig. 3. Example of UI derivation from a block of role-enriched business process model

take only one of these patterns as an example to illustrate the abstraction and aggregation (see Step 1 of Fig. 3). Step 1 indicates a scenario of arbitrary number of tasks in a Parallel-A block, where the tasks in the shaded area are from arbitrary number branches. According to the characteristics of this pattern, the execution of tasks in one branch will not affect tasks in other branches. Therefore, these tasks in the shaded area are aggregated to one single abstract node $ABCDEFG$. As a result, this abstract node and the remaining tasks are still in strict-order parallel.

5   Data Relationship Derivation

Here we refer to data relationships as the temporal relations between data items. A data item is associated with a single attribute operation in a task. By deriving the data relationships, the logic of control flow relations and data flow relations can be obtained, which are the foundation to build UI models. In this section, we will firstly introduce the syntax and symbols of data relationships, then demonstrate the derivation of data relationships.

5.1   Syntax and Symbols of Data Relations

In this section, we will introduce the syntax and symbols of the data relationships, which are demonstrated in Fig. 4.

The data relations can be summarized as three categories. The first category describes the existence of data sets, including **Co-existence**, and **Exclusiveness**. The second category describes the temporal ordering between data sets, including **Strict-order**, **Free-order**, **No-order**, **Parallel-A**, **Parallel-B**, **Conditional**, and **Loop**. The third category describes whether or not a data set can be separated, which is **Non-separation**.

5.2   Derivation Principles

In order to derive the data relationships for each user role, we will discuss the principles to derive the temporal relations of data items from the abstracted and aggregated business process model. The derivation principles contain two aspects: derivation based on control flow patterns and derivation based on the data flow patterns. Both the control flow patterns and the data flow patterns are from the abstracted and aggregated business process models.

We specify the derivation principles of data relationships as constraints and recommendations. Constraints are the rules which must be followed by UI designers during their UI design. These constraints will ensure the correct implementation of tasks. The recommendations are the rules which are not compulsory for

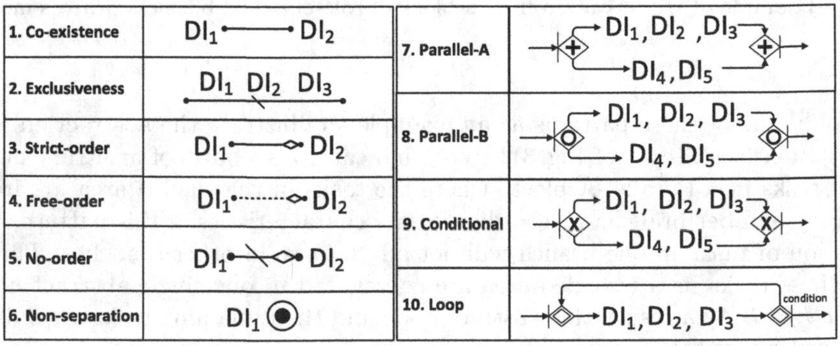

Fig. 4. Syntax and symbols of data relationships

the UI designers, which means that if these recommendations are not obeyed by UI designers, the correct implementation of tasks will not be affected. However, these recommendations will facilitate UI design to better reflect the logic of the underlying role-enriched business process model.

Step 2 of Fig. 3 is an example of deriving data relationships from *Parallel-A*. In this scenario, all the attribute operations within an individual task constitute a set of data items. As must-followed relationships, (1) the data item sets DI_H, DI_I, DI_J, which come from the tasks H, I, J of a single branch, must keep the same order as the tasks H, I, J in this branch; (2) the data item sets DI_K, DI_L, DI_M, DI_N should also keep the same order as the tasks K, L, M, N in their corresponding branch; (3) DI_H, DI_I, DI_J and DI_K, DI_L, DI_M, DI_N, which are from the same branches respectively, must remain in *Parallel-A*. We recommend: (1) DI_H, DI_I, DI_J, corresponding to the tasks H, I, J of a single branch, do not co-exist with each other; (2) so do the data item sets DI_K, DI_L, DI_M, DI_N; (3) the abstraction node $Agg_{ABCDEFG}$ and DI_H, DI_I, DI_J, and DI_K, DI_L, DI_M, DI_N must remain in *Parallel-A*.

6   UI Derivation

UIs of business processes are the media to realize the interaction between end users and BPMSs. Through the UIs of business processes, end users can read the data from systems and provide input data to systems. In this section, we will provide the specification of UI model, then specify UI derivation rules.

6.1   UI Model

A UI model comprises two layers: the relations between UI containers (containers), and the relationships between data items within a UI container. A container represents a UI which holds the maximum amount of data items to be operated by a user role. Basically, the container is a set of data items that can be directly implemented as a single web page containing the data items at its upper limit. At this situation, the entire content of the container will be shown to end users at a time. Alternatively, the UI designers can also divide this container into several sub-containers and implement these sub-containers as several web pages, which is a customized design. Inside a container, a data item indicates an attribute to be operated by end users within a single web page. Two aspects need to be specified for a data item: (1) **Access type** for each data item within a container, we need to specify the access type (*read/write*). (2) **Default/Compulsory/Optional** for every data item within a single container, we need to specify whether it is default, compulsory or optional. Only if all the default and compulsory data items are completed, can this container be completed.

Specification 1: UI Container. A UI container is denoted as $con = (DI,$ *access, prio, order*), where:

- $DI = \{di_1, di_2, ..., di_n\}$ is a finite set of data items.
- *access*: $DI \rightarrow \{read, write\}$ is a mapping function to assign the access type to each data item.
- *prio*: $DI \rightarrow \{default, compulsory, optional\}$ is a mapping function to assign the priority level to each data item, which decides whether a data item is default, compulsory or optional to execute.
- $ORDER$: $DI \times DI$ is to describe execution orders between data items within a UI container.

Specification 2: UI Model. A UI model for user role r is denoted as $UIM^r = (CON^r, con_s^r, con_e^r, G_{UI}^r, type_G^{UI,r}, CF_{ex}^{UI,r}, CF_{im}^{UI,r})$, where:

- CON^r is a finite set of UI containers.
- con_s^r and con_e^r are starting container and ending UI container, respectively.
- $G_{UI}^r = G_{in}^{UI,r} \cup G_{out}^{UI,r}$ is a finite set of gateways.
- $type_G^{UI,r}$: $G_{UI}^r \rightarrow \{strict\text{-}order\ sequential,\ free\text{-}order\ sequential,\ parallel\text{-}A,\ parallel\text{-}B,\ conditional,\ strict\text{-}loop,\ free\text{-}loop\}$ is a mapping function used to indicate the corresponding type of each gateway.
- $CF_{ex}^{UI,r}$, and $CF_{im}^{UI,r}$ are explicit and implicit control flow relations, respectively.

6.2   Derivation Rules

The UI model is derived based on the data relationships. The derivation of UI model contains two steps: (1) deriving the UI containers and the control flow relations between them; (2) deriving the data items and their relationships within each container. If two sets of data items DI_1 and DI_2 are in the data relationship *Co-existence, Exclusiveness, Free-order, Parallel-A, Parallel-B, Conditional*, or *Loop*, DI_1 and DI_2 must form two separate containers con_1 and con_2, meanwhile con_1 and con_2 should follow the same data relationship between f DI_1 and DI_2. If DI_1 and DI_2 are in *Strict-order, No-order* and *Non-separation*, these two data item sets should be in the same container, and the data items in this container should keep the same data relationships as that between DI_1 and DI_2.

We use the data relationships derived in Sect. 5 to demonstrate this UI derivation. In Step 3 of Fig. 3, data items DI_H, DI_I, DI_J form a UI container con_{HIJ}; DI_K, DI_L, DI_M, DI_N form a UI container con_{KLMN}; and abstract node $Agg_{ABCDEFG}$ will form a UI container $con_{Agg_{ABCDEFG}}$. In con_{HIJ}, DI_H, DI_I, DI_J remain the same order as in the data relationships; in con_{KLMN}, DI_K, DI_L, DI_M, DI_N also remain the same order as that in the data relationships.

7   Related Work

J. Kolb et al. [3] proposed a five-step method to generate UI components for process-aware information systems, in which a series of elementary and complex patterns were identified for UI generation. UI components were derived for each participating role. The proposed UI transformation method supported the propagation of changes in the business process. As the operated data inside each task have not been specified in their business process model, the operation flows between data items within a UI could not be derived, and each data item could not be specified with options: default, compulsory or optional.

K. Sousa et al. [4] introduced a four-step approach to derive UIs from business processes, including process modelling, task derivation, task refinement, and UI model derivation. This approach provided a solution to build up an abstract UI logic. The derived UI logic could not provide a customized result for each role; and within a UI, data items could not be specified with options: default, compulsory or optional.

V. Kunzle et al. in [5] proposed an object-aware approach for process modelling to derive UIs, which focused on the evolution of data objects. Tasks were modelled by considering both data object behaviour and interactions. The operation flows between UIs could not be generated for each role. Besides, inside each UI, the execution orders between data items were not derived.

Artifact-centric modelling is a paradigm to build business process models, which focus on data modelling of business process [6]. S. Yongchareon et al. [7] utilized this modelling approach to realize the automatic UI derivation. They derived UI flow models for each involved user role based on the behavioral and informational features of process models. Besides, the IBM team [8] developed a Barcelona prototype, which supported UI components derivation from artefact-centric process models. However, it failed to generate the operation flows between data items within a UI and the priority level for each data item within an UI could not be specified.

8   Conclusion and Future Work

This paper reports our current research progress on UI modelling and derivation for BPs. A framework has been proposed to support UI analysing, generating, and updating. A role-enriched business process model has been proposed as the foundation to identify control flow patterns and data flow patterns. These identified patterns are used to analyse the details of task and data relationships associated with UI derivation. A UI model, UI constraints, and recommendations for UI derivation have been presented and discussed. This work is still at its early stage. Detailed rules of constraints and recommendations will be specified in our future work.

Acknowledgments. This work is supported by the Australian Research Council Linkage Project (LP120200231) and the China Scholarship Council.

References

1. Weske, M.: Business Process Management: Concepts, Languages, Architectures. Springer Science & Business Media, Heidelberg (2012)
2. Van Der Aalst, W., Van Hee, K.M.: Workflow Management: Models, Methods And Systems. MIT Press, Cambridge (2004)
3. Kolb, J., Hübner, P., Reichert, M.: Model-driven user interface generation and adaptation in process-aware information systems (2012)
4. Sousa, K., Mendonça, H., Vanderdonckt, J., Rogier, E., Vandermeulen, J.: User interface derivation from business processes: a model-driven approach for organizational engineering. In: Proceedings of the 2008 ACM symposium on Applied Computing, pp. 553–560. ACM (2008)
5. Künzle, V., Reichert, M.: Philharmonicflows: towards a framework for object-aware process management. J. Softw. Maint. Evol. Res. Pract. **23**, 205–244 (2011)
6. Cohn, D., Hull, R.: Business artifacts: a data-centric approach to modeling business operations and processes. Bull. IEEE Comput. Soc. Techn. Comm. Data Eng. **32**, 3–9 (2009)
7. Yongchareon, S., Liu, C., Zhao, X., Xu, J.: An artifact-centric approach to generating web-based business process driven user interfaces. In: Chen, L., Triantafillou, P., Suel, T. (eds.) WISE 2010. LNCS, vol. 6488, pp. 419–427. Springer, Heidelberg (2010)
8. Heath III, F.T., Boaz, D., Gupta, M., Vaculín, R., Sun, Y., Hull, R., Limonad, L.: Barcelona: a design and runtime environment for declarative artifact-centric BPM. In: Basu, S., Pautasso, C., Zhang, L., Fu, X. (eds.) ICSOC 2013. LNCS, vol. 8274, pp. 705–709. Springer, Heidelberg (2013)

Process Modeling Leveraged by Workflow Structure and Running Logs Analysis

Fei Yu[1,2], Lipeng Guo[1,2], and Liang Zhang[1,2(✉)]

[1] School of Computer Science, Fudan University, Shanghai, China
{13210240041,10110240002,lzhang}@fudan.edu.cn
[2] Shanghai Key Laboratory of Data Science, Fudan University, Shanghai, China

Abstract. The reality of big data opens up a new world for business process modeling. Omnipresent cases and workflow logs are getting accessible, which implies the chance to exploit important patterns hidden in them so as to cut down the modeling cost or to improve the quality of process models. To take the full advantage of big data in process-aware systems (PASs), we propose a novel business process modeling technique that leverages the modeling by cases and workflow logs analysis. It uses the average perceptron to analyze both of existing process structure of cases and co-occurrence relation of activities in workflow logs. In contrast to traditional manual efforts, it improves the performance significantly by recommending proved working patterns. Comparing to recent process mining strategies, it serves the modeling online with meaningful process segments. We evaluate our approach against a synthesis dataset (100 processes and 10,000 log items generated by the plugin *PLG* in ProM) and real data from public business processes (77 processes in the package *Paul Fisher workflows for benchmarks PR and CA2* from the website *myExperiment*). The study reveals that 9.46 % improvement in precision can be gained by considering both case structure and log items in contrast to the structure only, or 5.94 % gaining in contrast to mere logs. Our evaluation validates the effectiveness of the proposed technique and efficiency when we applying it on real modeling scenarios.

Keywords: Business process modeling · Workflow logs · Average perceptron

1 Introduction

Process design phase would be inefficient if every time a company engages in modeling and redesigning its process, it did so "from scratch" without the consideration of how other companies or other designers in its own company perform similar processes. The main challenge is how to improve the efficiency of business process modeling, instead of modeling by human-based experience. First, designers are not always equivalent to domain experts. The diversity of demand from customer increases and designers often fall into dilemma because of lack of domain knowledge. Second, the quality of models is different from one designer to another even if solving the same problem. The overall quality of the whole system is instable as each variable alters which brings a lot of uncertainty.

Effective and efficient recommendation technique for activities in business process is useful and important for the following three reasons. First, it can speed up the

J. Cao et al. (Eds.): PAS 2015, CCIS 602, pp. 29–39, 2016.
DOI: 10.1007/978-981-10-1019-4_3

workflow construction process by reducing the development time. Second, it can provide a guidance for choosing the mostly likely activity and, therefore, minimizing the errors that are possibly made in the business process construction. Third, it help to uniform, reuse and process integration later.

Some existing approaches have fasten the design phase. Process mining [2, 4] succeeds in mining and obtaining business process models from running logs. However, they are always studied as a whole while sometimes only some parts of the model need to be considered. Pattern discovery strategies [5, 6] is past-knowledge-oriented and has achieved apparent success in extracting business patterns and process schema. Unfortunately, it can do little in real application scenarios.

In order to cut down the process modeling cost or improve the quality of process models, we propose a business process modeling technique using the average perceptron to analyze both the existing process structure of cases and co-occurrence relation of activities in workflow logs. As a discriminative model, compared with the generative model, average perceptron does not need to face the feature independence assumptions and it is suitable for high-dimensional sparse discrete feature space. So it is a general model to utilize more features to enhance the activity recommendation effect. When the designer design the process, we suggest the next possible activities as a short ranking lists of recommended activities. In order to test the validity of our approach, we evaluate it against a synthesis dataset and real data from public business processes in the domain of bioinformatics.

The rest of the paper is organized as follows. Section 2 deals with the model definition and presents an example that illustrates our approach. Section 3 explains the detail of this recommendation approach. Section 4 presents the evaluation implementation and experimental results. Section 5 discusses related work. Section 6 concludes the paper and discusses the future work.

2 Fundamental Overview

In this section, we present the preliminary about running logs, business process and the neighborhood context that are captured from them. We firstly present some definitions related to business process and illustrate a motivating example (Sect. 3.1). Then, we present why and how to make use of running logs (Sect. 3.2), and the manner to recommend activity based-on neighborhood context (Sect. 3.3).

2.1 Definition of Activity and Business Process

Without loss of generality, we select and use BPMN notations in our approach as it is one of the most popular business process modeling language [7], which are also easily mapped to equivalent notations in other business process modeling and workflow languages. In our work, we focus on nodes and the connections to their neighbors, and treat gateways the same way as activities to compute and recommend.

Let A_P be the set of all the nodes such as activities and gateways, and C_P be the set of connection links, which collects all the business process cases BP from a business

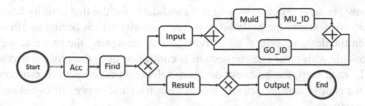

Fig. 1. A bioinformatics process model in BPMN notation

process library PL. For example, in Fig. 1, here is a sample business process BP_1 in its process library PL_1 (PL_1 contains all these such processes in the same domain). $A_{P1} = \{'a_i'\}(i = 1 \ldots n)$, n is the number of nodes in PL. For example, $A_{P1} = \{$'Start', 'Acc', 'Find', 'Muid', 'Mu_ID'…'End'$\}$, n = 14 which consists of 10 activities and 4 gateways, from which 'AND-join', 'XOR-join', 'AND-split' and 'XOR-split' are gateways and are the same with others. $C_p = \{$'Acc $\rightarrow$ Find', 'Find $\rightarrow$ XOR-split', 'XOR-split $\rightarrow$ Input', 'AND-split $\rightarrow$ Muid' … 'XOR-join $\rightarrow$ Query'$\}$, each element is a two-tuple which are the two endpoints of each link. It denotes the set of finite sequences over A_p.

2.2   The Neighborhood Relationship Extracted from Logs

Besides capturing the neighborhood contexts from existing business processes, we also discover from business process logs. These logs back up not only the business execution but also the knowledge related to the a-priori business process models. Our goal is to extract information from these logs in order to exploit the hidden knowledge that may be helpful for the business process recommendation.

$A*$ denotes the set of finite sequences over A_P and $\sigma = a_1a_2 \ldots a_n \in A*$ is a log trace. $L \in \mathcal{P}(A*)$ is a business process log. Taking the business process above as example, there is the number n_1 of log trace {Start $\rightarrow$ Acc $\rightarrow$ Find $\rightarrow$ XOR-split $\rightarrow$ Input $\rightarrow$ AND-split $\rightarrow$ Muid $\rightarrow$ MUID $\rightarrow$ GO_ID $\rightarrow$ AND-join $\rightarrow$ XOR-join $\rightarrow$ Output $\rightarrow$ End}, the number n_2 of log trace {Start $\rightarrow$ Acc $\rightarrow$ Find $\rightarrow$ XOR-split $\rightarrow$ Input $\rightarrow$ AND-split $\rightarrow$ GO_ID $\rightarrow$ Muid $\rightarrow$ MUID $\rightarrow$ AND-join $\rightarrow$ XOR-join $\rightarrow$ Output $\rightarrow$ End}, and the number n_3 of {Start $\rightarrow$ Acc $\rightarrow$ Find $\rightarrow$ XOR-split $\rightarrow$ Result $\rightarrow$ XOR-join $\rightarrow$ Output $\rightarrow$ End}. By extracting these process logs, we capture the execution orders between a_i in A_p. If $n_1 > n_2 > n_3$, it means the possibility of recommendation of 'Input' after 'XOR-split' is higher compared to 'Result'.

2.3   Recommendation Manner

Context is defined as a business process fragment around an activity, including the associated activity and connection flows connecting it and its neighbors. Based on the similarity values, we present for the business designer k activities that have highest similarity values. K is the size of the recommending list and user can change it as user's customization. Recommending the whole process can make the designer confused as other solutions in related works, especially in case of large-size business processes.

For a new business process, we suggest candidates for the first activity based on the process's description and the probability of each activity which occurs as after the start event. When the designer selects an activity to start designing the process, we suggest the next possible activities. The suggestion is continued until the process is completely designed. In each step, the business designer selects an activity from our recommendation list (or choose another activity beyond our list) and creates links between it and previous activities.

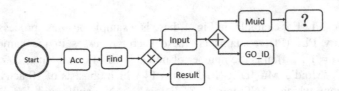

Fig. 2. An example when the Fig. 1 business process is creating ongoing

3   Recommendation Approach

We propose to use a discriminative model named average perceptron [3] to train and predict the recommended activity. The reason to apply average perceptron is: (1) As linear architectures, it perform better than non-linear ones and other algorithm especially in high-dimensional sparse discrete feature space. So it is suitable exactly for solving our problem since our mapping table towards feature library is very sparse. (2) Compared to the generative model, such as Naïve Bayes, it does not need to face the feature independence assumptions. So it is a general model to utilize more features to enhance the activity recommendation effect.

Algorithm 1: Train

```
Input: workflow_dataset
Output: Average Perceptron Model
Initialize: w  = 0, w_avg = 0
 1: train_dataset = [];
 2: for each workflow in workflow_dataset do
 3:     for each activity in workflow do
 4:         y = node.name;
 5:         x = get_feature(node, workflow);
 6:         train_dataset.add(x, y);
 7:     end for
 8: end for
 9: for t = 1 to T do
10:     for each(x,y)in train_dataset do //dataset number is N
11:         ŷ = argmax_z wᵀφ(x, z);
12:         if y != ŷ then
13:             w = w + α · (φ(x,y) − φ(x,ŷ));
14:         end if
15:         w_avg = w_avg + w;
16:     end for
17: end for
18: return w_avg/(N ∗ T);
```

The train method is as following (Algorithm 1). First, we need to transform the contexts and the activities in business process to a feature and a label, which is named (x, y) as a training instance. Then we can build the instance list of training data sets (X, Y). Secondly, we train the model by average perceptron model. At last, we get the model to predict the recommended activities in the test method.

3.1 Feature Extraction

First, we need to extract the feature to build the instance list (X, Y). For every instance (x, y) in the instance list, x stands for the features extracted from the neighborhood context around the recommending activity in its business process structure, y stands for the activity ID which we should predict to recommend.

Take the example above as an example (Fig. 1), if the business process is now creating ongoing as the activities of Fig. 2, then we need to recommend which one is next to 'Muid'. At this moment, x = { 'previous is Muid', 'penult is And-split', 'penult's next is Go_ID'...} and y = {MU_ID}. For the logs library, we can also use this feature extraction method. The main difference between log and process is that the log will always be a serial sequence process.

3.2 Training Procedure

We use the average perceptron algorithm to train the model parameters (Algorithm 1). The average strategy is used to avoid the overfitting problem.

Given an example (x, y), we need to predict the label $\hat{y}$ with the highest score.

$$\hat{y} = argmax_z \, w^T \varphi(x, z) \tag{1}$$

where w is the parameter of score function, and $\varphi(x, z)$ is the feature vector consisting of lots of overlapping features, which is the chief benefit of discriminative model. As can be seen from Eq. (1), features sparsity doesn't affect our algorithm.

For the training data sets, given an example (x, y), we compare the y and $\hat{y}$ as the hinge-loss function. If $\hat{y} = y$, it proves that the prediction is correct, we do not need to change the weight w. If $\hat{y} \neq y$, it proves that the prediction is incorrect, we need to update the weight w, it can be calculated by

$$w = w + \alpha \cdot (\varphi(x, y) - \varphi(x, \hat{y})) \tag{2}$$

where α is the learning rate, it represents the updated step size in the range from 0.0 to 1.0. And it is generally chosen as 0.1.

When the algorithm trains the whole training dataset for T iterations, it will stop and return the model which we have already trained. T is a hyper-parameter, and we use cross-validation to minimum the training error and then find the well-designed T = 14. We use the average weight w_{avg} to record the sum of the weight w, and return $w_{avg}/(N * T)$ as the model. w_{avg} is better for predicting than the weight w in practice, and the divisor $N * T$ is for reducing the size of weights in order to avoid predicting overflow problem.

3.3   Testing Procedure

As the Algorithm 2, when we get the workflow dataset, we make the model predict the recommended activity just like the way that the workflow designer do. When the designer make a connection link, our model will extract the context feature and recommend the activity which the link connected to. The designer may choose to accept our recommendation or reject our recommendation (choose another activity beyond our list), then he will make another connection link as next step. He will continue until he finishes the whole business process. So we will test our model on the test dataset as the same way. We will calculate the recommendation accuracy for all the activities in the workflow and take their previous context as the features.

When we only recommend one activity, we evaluate the recommended activity as "good" only if the recommended activity is the same as the real activity. We can also produce the top-k recommended activity list for the designer to choose, and we consider the recommended activity is "good" when the real activity is included in the top-k recommended activity list.

Algorithm 2: Test

```
Input: workflow_dataset
Output: model predict accuracy
 1: sum=0; //number of all the activity
 2: good=0;//number of activities whose predication is right
 3: for each workflow in workflow_dataset do
 4:    for each activity in workflow do
 5:        y = activity.name;
 6:        x = get_feature(activity, workflow);
 7:        ŷ = argmax_z w^T φ(x, z);
 8:      sum += 1;
 9:        if y == ŷ then
10:            good += 1;
11:      end if
12:    end for
13: end for
14: return good/sum;
```

---

4   Evaluation

This section presents the implementation to realize our recommendation techniques, and the experiments to evaluate the effectiveness of our solutions. Besides, we analyze the parameters that impact on the recommendation quality.

4.1   Datasets

We evaluate on both real world dataset and simulation dataset to show the effectiveness and scalability of our method. The statistics in details is given in Table 1. As the real-world running logs dataset corresponding to the certain process dataset is difficult to obtain, we use a large scale of simulation dataset to simulate the log and process to test our method by combining both of them, as the same way as other related work [8, 9].

Real world dataset is downloaded from the *myexperiments* website.[1] They are 77 processes of *Paul Fisher* workflows for *benchmarks PR and CA2* in the domain of bioinformatics. For simulation dataset, synthesis is generated by *Process Log Generator* (PLG) [10], a plug-in for ProM framework which enables to create random BPMN models from common workflow patterns and to simulate execution of these processes (100 logs is generated for each process, so 10,000 logs totally). For business process models, we use PLG to do customization by changing basic pattern percentages: loop percentage, single activity percentage, sequence percentage, AND split-join percentage, XOR split-join percentage. For running logs, we choose noise level to generate little noise log records throughout execution to simulate the reality. For the real world dataset and the simulation dataset, we choose 80 % for training and 20 % for testing.

Table 1. Details of the dataset

	No. of process	No. of all activities	No. of gateway	No. of distinct activities	Gateway percentage
Real world	77	1089	348	269	31.95 %
Simulation	100	1857	973	41	52.39 %

4.2 Experiments

To situate the performance of our model, we first compare between the real world dataset and the simulation dataset (Fig. 3). We notice that the gap of accuracy between the two dataset in top-1 is very small, when the simulation is 49.18 % and the real world dataset is 38.42 %. However, as the value of k increases, the gap of accuracy also increases. The reason is that our real world dataset contains 269 distinct activities but only 1089 in the business process, meanwhile the simulation workflow dataset contains 41 distinct activities but 1857 totally. The sparsity of data makes the accuracy of this real world dataset maybe less than the simulation one. But we should also notice that our method can perform well in the top-k recommendation. The accuracy is 53.69 % when recommending top-6 activities, which means that the process designer can often get what he wants. Compared to the random recommendation which may achieve less than 1 % accuracy, thus our method is reliable for actual real recommendation.

We use the simulation process and log dataset to show the effectiveness of the log dataset. We use a mixture strategy to combine the workflow model and the log model.

$$\hat{y} = argmax_z \; \alpha w_{flow}^T \varphi_{flow}(x, z) + (1 - \alpha) w_{log}^T \varphi_{log}(x, z) \qquad (3)$$

where $\alpha \in [0, 1]$ is a hyper parameter. If $\alpha = 1$, it means that we only use the business process models to recommend as our baseline model to compare. If $\alpha = 0$, it means that we only use the log. $w_{flow}^T \varphi_{flow}(x, z)$. and $w_{log}^T \varphi_{log}(x, z)$ are the scores given by business process model and log model respectively.

---

[1] http://www.myexperiment.org/packs.

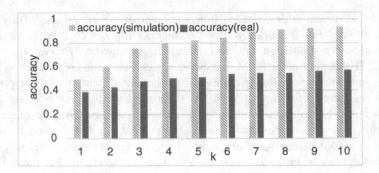

Fig. 3. Accuracy on the real world dataset and the simulation dataset in process

Figure 4 shows the comparison among log model ($\alpha = 0$), process model ($\alpha = 1$) and combined model ($0 < \alpha < 1$). Compared by process model and combined model, though the gap of accuracy in the top-1 may be small, the combined model can achieve significant improvement when k is large. In reality, we will always recommend top-6 activities to the designer. Because when the k is small, the designer may have no choice for the recommendation, and when the k is very large the designer may be confused. When the k is 6, the combined model will achieve 93.51 % (+9.46 %) and the process model will only achieve 84.05 % and the log model will only achieve 87.57 %. The result shows the effectiveness and contribution of the log dataset. The reason behind this is that the log dataset can reflect the probability that the log trace go through the recommended activity, and this information cannot be included in the workflow structure. As a result, if we can get the log library dataset corresponding to the business process library dataset in reality, we can use the combined model to improve the accuracy and enhance the recommendation effects. The time complexity doesn't change but the training time increases as the scale of features increases.

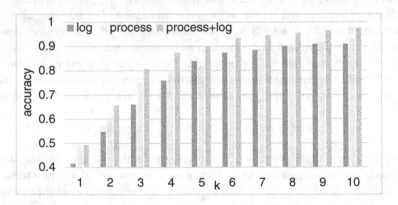

Fig. 4. Accuracy on process ($\alpha = 1$), log ($\alpha = 0$) and combined model

5 Related Work

Some existing approaches [11–13] target to fasten the design phase by retrieving similar process and their fragments to match the current designed process from repositories. They are proposed either to rank existing business process models for similarity search [1, 11], or to measure the similarity between them [12, 14] for creating new process models. In our approach, we focus partially on the business process and take into account only the activity neighborhood context for recommendations instead of matching the whole business process.

Process mining [2, 4], also known in the industry as Automated Business Process Discovery (ABPD), mines and obtains business process models based on logs analysis. It can discover process, control, data, organizational, and social structures from event logs, but they are always studied as a whole while sometimes only some parts of the model need to be considered.

Pattern discovery [5, 6] is a past-knowledge-oriented technique and has achieved apparent success in extracting business pattern and process schema. However, it can do little in real application scenarios by far.

Dijkman et al. [17] used Levenshtein distance to compare the activity labels; graph edit distance and vector space model to determine the similarity between business process structures. They also proposed the ICoP framework [16] to identify the match between parts of process models using these metrics. Different from them, we focused on activity neighborhood contexts extracted from both business process and log library, instead of matching activity labels or matching virtual documents.

Cao et al. [15] proposed Near Neighbor and Maximal Subgraph First (NMSF) based and processed warping matrix as their business process recommendation technique, using graph mining technique to extract the process patterns to judge the equivalence between the reference process and to process patterns for determining the candidate node sets for recommendation. But the former only supports accurate recommendation, and the latter must be computed on line leading to huge time costing. Our approach can support fuzzy recommendation and doesn't have the excessive consumption time.

Zhang et al. [8] proposed a workflow recommendation technique called Flow-Recommender. FlowRecommender needs predefined patterns which register offline to ensure a highly efficient online workflow recommendation. It isn't necessary to provide extra information in our recommendation.

Nguyen et al. [7, 9] have done a lot of valuable work about assisting business process design, mainly using k-zones after transforming the graph of work-flow based on graph theory. As a kind of generative models, it is limited to the assumption of independence between features. As a discriminative model, our approach can overcome that limitation and fully exploit the relationship underlying.

6 Conclusion and Discussion

We propose to analysis process and execution logs to provide modeling suggestions to developers depending on their last modeling action. As discriminative model, instead of generative models, average perceptron is able to combine workflow structure and logs, and makes full use of these abundant information with practical value as extracted feature.

For a selected activity to recommend, we sort activities in descending order of similarity and pick up top-k activity for recommendation to designers. Experimental results showed that our approach was accurate and has good performance according to obtained precision values. Our approach, which is off-line learning, can be further improved to achieve more individuation and customization in the future work. Considered users' early choice as feature of a new instance, the model can be modified so as to become more precise and suitable to the users' personal preferences.

Acknowledgments. The work is partially supported by NSFC (No. 60873115), Shanghai Science and Technology Development Funds (No. 13dz2260200 & No. 13511504300), and National Hi-Tech. Project (2012AA02A602).

References

1. Roy Chowdhury, S., Daniel, F., Casati, F.: Recommendation and weaving of reusable mashup model patterns for assisted development. ACM Trans. Internet Technol. **14**(2–3), 21:1–21:23 (2014)
2. Van der Aalst, W., Weijters, T., et al.: Workflow mining: discovering process models from event logs. IEEE Trans. Knowl. Data Eng. **16**(9), 1128–1142 (2004)
3. Crammer, K., Dekel, O., Keshet, J., et al.: Online passive-aggressive algorithms. J. Mach. Learn. Res. **7**, 551–585 (2006)
4. Yang, L.Q., Kang, G.S., Guo, L.P., Zhang, L., Zhang, X.N., Gao, X.: Process mining approach for diverse application environments. J. Softw. **26**(3), 550–561 (2015). (in Chinese)
5. van der Aalst, W.M., Song, M.S.: Mining social networks: uncovering interaction patterns in business processes. In: Desel, J., Pernici, B., Weske, M. (eds.) BPM 2004. LNCS, vol. 3080, pp. 244–260. Springer, Heidelberg (2004)
6. Lv, Y., Lei, J.-K., Yang, L.-Q., Zhang, L.: BPDetector: a extensible business pattern detector framework. SOARingLab TR #20150802, Fudan Univesity, China (2015, inpress in Chinese)
7. Nguyen, N.C., et. al.: Service recommendation for individual and process use. Institut National des Télécommunications (2012)
8. Zhang, J., Liu, Q., Xu, K.: FlowRecommender: a workflow recommendation technique for process provenance. In: Proceedings of the 8th Australasian Data Mining Conference, vol. 101, pp. 55–61. Australian Computer Society, Inc. (2009)
9. Chan, N.N., Gaaloul, W., Tata, S.: Context-based service recommendation for assisting business process design. In: Huemer, C., Setzer, T. (eds.) EC-Web 2011. LNBIP, vol. 85, pp. 39–51. Springer, Heidelberg (2011)

10. Burattin, A., Sperduti, A.: PLG: a framework for the generation of business process models and their execution logs. In: Muehlen, Mz, Su, J. (eds.) BPM 2010 Workshops. LNBIP, vol. 66, pp. 214–219. Springer, Heidelberg (2011)
11. Yan, Z., Dijkman, R., Grefen, P.: Fast business process similarity search with feature-based similarity estimation. In: Meersman, R., Dillon, T.S., Herrero, P. (eds.) OTM 2010. LNCS, vol. 6426, pp. 60–77. Springer, Heidelberg (2010)
12. Zha, H., Wang, J., Wen, L., et al.: A workflow net similarity measure based on transition adjacency relations. Comput. Ind. **61**(5), 463–471 (2010)
13. Wen, L., Wang, J., Sun, J.: Detecting implicit dependencies between tasks from event logs. In: Zhou, X., Li, J., Shen, H.T., Kitsuregawa, M., Zhang, Y. (eds.) APWeb 2006. LNCS, vol. 3841, pp. 591–603. Springer, Heidelberg (2006)
14. Jin, T., Wang, J., Wen, L.: Efficient retrieval of similar business process models based on structure. In: Meersman, R., et al. (eds.) OTM 2011, Part I. LNCS, vol. 7044, pp. 56–63. Springer, Heidelberg (2011)
15. Cao, B., Yin, J., Deng, S., et al.: A near neighbour and maximal subgraverage perceptron first based business process recommendation technique. Chin. J. Comput. **36**(2), 263–274 (2013)
16. Weidlich, M., Dijkman, R., Mendling, J.: The ICoP framework: identification of correspondences between process models. In: Pernici, B. (ed.) CAiSE 2010. LNCS, vol. 6051, pp. 483–498. Springer, Heidelberg (2010)
17. Dijkman, R., Dumas, M., Van Dongen, B., et al.: Similarity of business process models: metrics and evaluation. Inf. Syst. **36**(2), 498–516 (2011)

CWFDesigner: A Tool for Supporting the Visual Modeling and Analysis of Cloud Workflow with Various Modeling Languages

Hua Huang[1,2], Rong Peng[1(✉)], and Zaiwen Feng[1]

[1] State Key Laboratory of Software Engineering, School of Computer, Wuhan University,
Wuhan 430072, China
{rongpeng,zwfeng}@whu.edu.cn
[2] School of Information Engineering, Jingdezhen Ceramic Institute,
Jingdezhen 333001, Jiangxi, China
jdz_hh@qq.com

Abstract. Process modeling is the key activity for users to apply cloud workflow service, and the retrieval, transformation and analysis of different format process models is the key technology for the realization of users' (enterprise tenants) internal business process integration and optimization. Most existing process modeling tools only support a specific modeling language and cannot achieve the retrieval, transformation and analysis of multi-format process repositories. In addition, these tools must be installed before using them. Therefore, we proposed a cloud workflow model (CWF), which does not depend on a specific modeling language and developed a tool (namely CWFDesigner) for supporting the visual modeling and transformation of cloud workflow with various modeling languages. Since this tool is deployed in a cloud computing environment, users can adopt the browser to build and analyze business process models without installing it. The actual practice shows that this tool is simple and easy to use and has good compatibility.

Keywords: Cloud workflow · Visual modeling · Process model transformation · Process model analysis

1 Introduction

Cloud workflow is a new application mode of workflow management systems deployed in the cloud computing environment. Cloud workflow systems can be widely used as platform software (or middleware services) to facilitate the usage of cloud services. Every tenant can design, configure and run his/her business processes on it [9]. In cloud workflow systems, process modeling is a key activity for users to apply cloud workflow service, and the retrieval, transformation and analysis between different format process models is the key technology for the realization of users' (enterprise tenants) internal business process integration and optimization [2, 10].

So far there already exist a variety of business process modeling language, such as, Petri-net, YAWL (Yet Another Workflow Language), BPMN (Business Process Modeling

© Springer Science+Business Media Singapore 2016
J. Cao et al. (Eds.): PAS 2015, CCIS 602, pp. 40–49, 2016.
DOI: 10.1007/978-981-10-1019-4_4

Notation), EPC (Event-driven Process Chain), UML AD (Unified Modeling Language Activity Diagram), BPEL (Business Process Execution Language) etc. [3, 4, 5]. For each kind of modeling language, the corresponding tool is developed to achieve the visual modeling of business processes [6]. However, most the existing process modeling tools must be installed. Meanwhile, they only support one specific modeling language and not support the transformation between different format process models. Although there are some tools, e.g., Apromore [7] and BeehiveZ [8], which support the modeling and analysis of different format processes, but Apromore does not support the automatically generating of the simulation processes and the direct transformation between different format process models (in Apromore platform, all process models are converted into Petri net models), and BeehiveZ is a Java application, which needs to be downloaded and installed.

In cloud workflow systems, different users may be familiar with different modeling language. If the design tool of cloud workflow only supports a specific modeling language, it may reduce the modeling efficiency of users that are familiar with other language, which will put negative effects on the popularization and application of cloud workflow. To address this problem, we analyzed the modeling rules and syntax characteristics of several usual modeling languages, e.g., Petri net, YAWL, BPMN, EPC, extracted their common characteristics and proposed a Cloud Workflow model (namely CWF) that does not depend on a specific modeling language. Based on CWF, we designed and implemented a tool (denoted as CWFDesigner) for supporting the visual modeling and analysis of cloud workflow. The formal definition of CWF, the system architecture of CWFDesigner and its key implementation technologies are presented as follow.

2   Model Definition

In cloud workflow systems, a number of different users can also design, configure, run their business processes. In the design and validation phase of users' workflows, to support different users to achieve process design and analysis with different modeling language, CWF must be an abstract model [8] with a higher virtualization. As described in [12], the topological structure of the process model constructed by any kind of language can be abstracted as a graph (namely Business Process Graph, BPG) that is composed of nodes and edges [2]. In order to enhance the description capability of BPG, we extract the basic modeling notation of four usual modeling languages: Petri net, YAWL, BPMN, EPC to extend the definition of the process node (without considering the resources and data attributes), then the extended process node is applied in the formal definition of CWF.

Definition 1 (Process Node):   A process node *t* (not including *Start* node and *End* node) is a 4-tuple, *t* = (*id, label, type, attr*), in which:

- *id* specifies the unique identifier of a process node.
- *Label* ∈ *L* specifies a specific label of a process node, where *L* is the set of labels.
- *type* ∈ *TP* specifies a specific type of a process node, where *TP* is the set of node types, which is dynamically generated according to the category of modeling

language (namely process type). If the process type is Petri-net, then $TP = \{$Transition, Place$\}$; If the process type is YAWL, then $TP = \{$Atomic Task, Compound Task, Multi-instance Atomic Task, Multi-instance Compound Task, Condition Node$\}$; If the process type is BPMN, then $TP = \{$Activity, Gateway, Event$\}$; If the process type is EPC, then $TP = \{$Function, Connector, Event$\}$.

- *Attr* $\in AT$ specifies a specific attribute of a process node, where AT is the set of node attributes, which is dynamically generated according to the node type (namely *type*). If *type* $\in \{$Atomic Task, Compound Task, Multi-instance Atomic Task, Multi-instance Compound Task$\}$, then $AT = \{$None-Split, None-Join, And-Split, OR-Split, XOR-Split, And-Join, OR-Join, XOR-Join$\}$; If *type* $\in \{$Gateway,Connector$\}$, then $AT = \{$And-Split, OR-Split, XOR-Split, And-Join, OR-Join, XOR-Join$\}$; If *type* $\in \{$Event$\}$, then $AT = \{$Intermediate Event, Message Event, Timer Event$\}$; Other else, $AT = $ null.

Definition 2 (Cloud Workflow Model, CWF): Cloud workflow model (CWF) is a 5-tuple: CWF = (UID, PID, PType, N, E), where:

- UID specifies the creator ID of a process model;
- PID specifies the unique identifier of a process model;
- PType $\in PT$ specifies the category of the modeling language adopted in a process model, where PT is the set of the categories of modeling languages supported by CWFDesigner, $PT = \{$Petri-net, YAWL, BPMN, EPC$\}$;
- N = $\{Start\} \cup \{End\} \cup T$, *Start* and *End* specify the start node and the end node respectively, T is the set of process nodes in a process model, $T = \{t_i | i = 1, 2, 3, ..., n\}$, t_i is the ith process node;
- E = $<Start \times T> \cup <T \times T> \cup <T \times End>$ is a set of flow relations (namely edges), which describes the execution orders between process nodes (namely tasks).

According to Definitions 1 and 2, cloud workflow model can meet the requirement of quickly building and analyzing different format business process models for different users that are familiar with different modeling language (e.g. Petri net, YAWL, BPMN, EPC). Namely, cloud workflow model is the basis for designing and implementing the CWFDesigner tool. To better understand the implementation process of CWFDesigner, its system architecture is presented below.

3   System Architecture

The main function of CWFDesigner is composed of four modules: model design, model retrieval, model transformation and model analysis, the system architecture of CWFDesigner is shown in Fig. 1.

In Fig. 1, the model design module is used for users to visually design their business process models and to automatically generate the simulation processes used in related experiments. The model retrieval module is used to achieve exact or similar query of process models based on users' demand with various process model formats. The model transformation module is used for users to export the target models transformed from the source models according to the corresponding transformation rules. The model analysis module is used to check or validate the models designed by users or the transformed target models according to the grammar rules of the corresponding modeling languages, and the feedback results are sent to users to modify and adjust the models. Except for this, the degree of similarity between different process models is calculated to do cluster analysis. All the related data of process models is stored in the process model repository of cloud database server through the a database access middleware. In order to more clearly understand the system architecture of CWFDesigner, the detail description of the model design module, the model retrieval module, the model transformation module and the model analysis module are presented below.

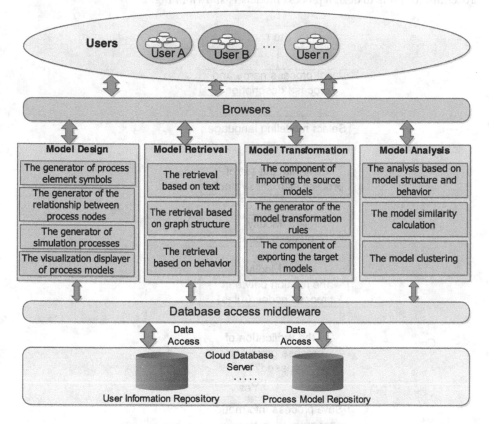

Fig. 1. The system architecture of CWFDesigner

3.1 Model Design

The model design module includes four main components: the generator of process element symbols, the generator of the relationship between process nodes, the generator of simulation processes, the visualization displayer of process models. The generator of process element symbols is used to generate the set of process element symbols according to the selected process modeling language, then the generated set of process element symbol is adopted for users to add or modify process nodes. The generator of simulation processes is used to automatically adjust or save the relationship between process nodes. The generator of simulation processes is used to automatically generate visual process models according to the process parameters reset by users, such as process type, the number of processes, the max number of nodes in a process model, the max in-degree or out-degree of a node, the control flow generating rules of the selected process type, and the labels randomly selected from the WordNet library. The visualization displayer of process models is used to obtain the process information (including the location and relationship of all nodes) and display them on web browsers. The detail procedure for users to design process models is shown in Fig. 2.

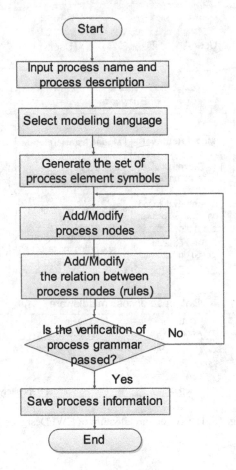

Fig. 2. The procedure of designing process models

3.2 Model Retrieval

The model retrieval module is composed of three main components: the retrieval based on text, the retrieval based on graph structure and the retrieval based on behavior. The component of the retrieval based on text is used to query the related process models in cloud workflow model repository according to the process name or process description input by users. The component of the retrieval based on graph structure is used to achieve the process model retrieval according to the process fragments input by users. The component of the retrieval based on behavior is used to query the most similar process models according to the process behavior description (such as the ordering relation between process task nodes). The detail procedure for users to query process models is shown in Fig. 3.

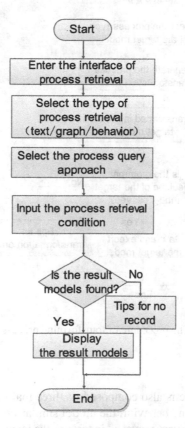

Fig. 3. The procedure of querying process models

3.3 Model Transformation

The model transformation module is composed of three main components: the component of importing the source models, the generator of the model transformation rules and the component of exporting the target models. The component of importing the source models is used for users to import the process models (XML files) that are needed to be transformed. The generator of the model transformation rules is used to generate the model transformation rules according to users' requirements and the control-flow pattern based transformation framework [10]. The component of exporting the target models is used to transform the imported source models into the target models according to the generated rules and export the XML file of the target models. The detail procedure for users to transform process models is shown in Fig. 4.

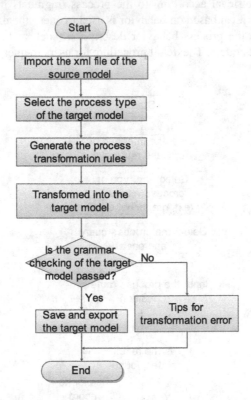

Fig. 4. The procedure of transforming process models

3.4 Model Analysis

The model analysis module is also composed of three main components: the analysis based on model structure and behavior, the model similarity calculation and the model clustering. Since Petri net is very powerful in term of the formal analysis and verification of process models, thus, all other process models are first converted into Petri net models before analyzing them in CWFDesigner. The component of the analysis based on model structure and behavior is used to generate the coverability tree and complete firing sequences of process models. Then the quantitative relationship between process nodes (e.g., the ordering relation with time constraints) is obtained to verify the correctness of the model structure and the rationality of the process behavior. The component of the model similarity calculation is used to calculate the similarity between different process models, including the node label semantic similarity, the process structure similarity and the process behavior similarity. The component of the model clustering is used to cluster the selected process model set according to the model similarity calculation results. The detail procedure for users to cluster process models is shown in Fig. 5.

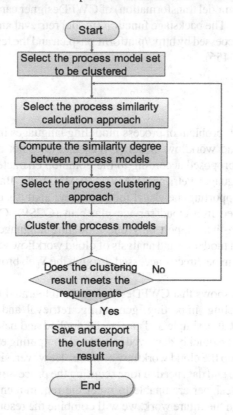

Fig. 5. The procedure of clustering process models

4   The Implementation of CWFDesigner

CWFDesigner is developed based on the .Net Framework 4.0 and the Microsoft Silverlight plug-in in C#. Silverlight is a powerful development tool for creating engaging, interactive user experiences for Web and mobile applications. Silverlight is a free plugin, powered by the .NET framework and compatible with multiple browsers, devices and operating systems, bringing a new level of interactivity wherever the Web works. Using Silverlight for their applications, developers can create applications with richness and interactivity out of reach of traditional web technologies while retaining the simple deployment and update model of web applications. In addition, in order to support non-chinese users to use the tool, we have added multi-language pack to the tool (the current version supporting Chinese and English) for them to switch the languages. Therefore, CWFDesigner has stronger interaction experience effect and good compatibility.

The prototype of CWFDesigner is deployed in the Ceramic Cloud Service Platform (CCSP for short) that can be accessed by http://www.pasp.cn, the related data of CWFDesigner is also stored in the cloud database server of CCSP. The front desk function (model design and model transformation) of CWFDesigner can be directly accessed by http://cbpm.pasp.cn. The backstage function (model retrieval and model analysis) of CWFDesigner can be accessed by http://platform.pasp.cn and the test user (user account: jdztygk, password: 123456).

5   Conclusion

To address the diversity problem of process modeling languages in the design phase of cloud workflow, a cloud workflow model (CWF) that does not depend on a specific modeling language is proposed according to the common characteristics of four kinds of usual modeling language: Petri net, YAWL, BPMN, EPC. Based on CWF, a tool (CWFDesigner) for supporting the visual modeling and analysis of cloud workflow is developed and deployed in a cloud service platform (CCSP). Compared with other process modeling tools that support only one modeling language, CWFDesigner not only supports the visual modeling and analysis of cloud workflow with various modeling languages, but also it can be directly accessed through the Web browsers without installing it.

The actual practice shows that CWFDesigner is an integrated tool that can achieve the visualization modeling, importing, generating, retrieval and analysis of various format process models. It is simple and easy to be applied and has good compatibility. Meanwhile, it can also be quickly deployed in a cloud computing environment and play a positive role to develop the cloud workflow systems. However, since CWF has not yet considered the resource and data needed for executing the processes, the process models constructed by CWFDesigner are unable to meet the requirement of cloud workflow execution. Therefore, in the future work, we will combine the resources and the process data to extend the cloud workflow model. Based on the extended cloud workflow model, we will further improve the CWFDesigner tool and integrate it into the cloud workflow execution engine.

Acknowledgments.   This work is supported by the National 973 Basic Research Program of China under Grant No. 2014CB340404, the National Natural Science Foundation of China under Grant Nos. 61170026,61373037 and 61100017, the National Science and Technology Ministry of China under Grant Nos. 2012BAH25F02 and 2013BAF02B01 and the Fundamental Research Funds for the Central Universities of China under Grant Nos. 2012211020203 and 2042014kf0237.

References

1. Kapuruge, M., Colman, A., Han, J.: Achieving multi-tenanted business processes in SaaS applications. In: Bouguettaya, A., Hauswirth, M., Liu, L. (eds.) WISE 2011. LNCS, vol. 6997, pp. 143–157. Springer, Heidelberg (2011)
2. Mendling, J., Reijers, H.A., van der Aalst, W.M.P.: Seven process modeling guide-lines (7PMG). Inf. Softw. Technol. **52**(2), 127–136 (2010)
3. Wohed, P., van der Aalst, W.M.P., Dumas, M., ter Hofstede, A.H.M., Russell, N.: On the suitability of BPMN for business process modelling. In: Dustdar, S., Fiadeiro, J.L., Sheth, A.P. (eds.) BPM 2006. LNCS, vol. 4102, pp. 161–176. Springer, Heidelberg (2006)
4. Van der Aalst, W.M.P.: Verification of workflow nets. In: Azéma, P., Balbo, G. (eds.) ICATPN 1997. LNCS, vol. 1248, pp. 407–426. Springer, Heidelberg (1997)
5. Van Der Aalst, W.M.P., Ter Hofstede, A.H.M.: YAWL: yet another workflow language. Inf. Syst. **30**(4), 245–275 (2005)
6. van Dongen, B.F., de Medeiros, A.K.A., Verbeek, H.M.W., Weijters, A.J.M.M., van der Aalst, W.M.P.: The ProM framework: a new era in process mining tool support. In: Ciardo, G., Darondeau, P. (eds.) ICATPN 2005. LNCS, vol. 3536, pp. 444–454. Springer, Heidelberg (2005)
7. La Rosa, M., et al.: M.: APROMORE: an advanced process model repository. Expert Syst. Appl. **38**(6), 7029–7040 (2011)
8. Jin, T., Wang, J., Wen, L.: Efficiently querying business process models with BeehiveZ. In: BPM (Demos) (2011)
9. Xue-zhi, C., Jian, C.: Cloud computing oriented workflow technology. J. Chin. Comput. Syst. **33**(1), 90–95 (2012). (in Chinese with English abstract)
10. Zhen-Lu, S., Li, Z., Ji-min, L.: CP-PMTT: a control-flow pattern based process model transformation tool. Comput. Integr. Manuf. Syst. **21**(2), 304–315 (2015)
11. He, X., Ma, Z., Zhang, Y., Shao, W.: Extending QVT relations for business process model transformation. J. Softw. **22**(2), 195–209 (2011)
12. Wang, J., Jin, T., Wong, R.K., Wen, L.: Querying business process model repositories. World Wide Web **17**, 427–454 (2014)

Process Data Analysis

Analysis of Minimum Workflow Resource Requirement

Jiacun Wang[1(✉)], Bill Tepfenhart[1], and Xiaoou Li[2]

[1] Department of Computer Science and Software Engineering,
Monmouth University, Monmouth, USA
{jwang, btepfenh}@monmouth.edu
[2] Department of Computation, CINVESTAV-IPN,
Mexico, D.F., Mexico
lixo@cinvestav.mx

Abstract. A workflow describes the flow of jobs of a business process. Executing a workflow requires resources. In many situations, business processes are constrained by scarce resources. Therefore, it is important to understand workflow resource requirement. In our previous work, we introduced resource oriented workflow nets (ROWN) and based on ROWN, an efficient algorithm for the analysis of the maximum resource requirement of a workflow (maxRR) was developed [11]. The maxRR is the minimum amount of resources that support workflow execution along every possible path. On the hand, when there is a resource shortage, it is important to find the minimum resource requirement (minRR), which is the minimum amount of resources that support workflow execution along at least one path. In this paper, we present an approach to analyzing the minRR.

Keywords: Workflows · Workflow nets · Petri nets · Resource requirements analysis

1 Introduction

This paper focuses on workflow resource requirements analysis. In many situations, business processes are constrained by scarce resources. The lack of resources can cause contention, the need for some tasks to wait for others to complete, which slows down the progress of workflow execution. This is particularly true in an emergency response system where large quantities of resources, including emergency responders, ambulances, medical care personnel, fire trucks, medications, food, clothing, etc., are required [12]. Often potential delays can be avoided or reduced by using resource analysis to identify ways in which tasks can be executed in parallel, in the most efficient way. A workflow with resource usage defined can help keep track of resource availability, disable the paths that are not executable, and present all executable paths, thus allowing the automatic selection of feasible path for system execution. General workflow resource patterns are introduced in [9].

Petri nets are a powerful tool in modeling and analyzing resources-constrained systems. Some excellent research results on resource allocation and deadlock avoidance

© Springer Science+Business Media Singapore 2016
J. Cao et al. (Eds.): PAS 2015, CCIS 602, pp. 53–66, 2016.
DOI: 10.1007/978-981-10-1019-4_5

are published. For example, workflow nets have been identified and widely used as a solid model of business processes [2, 5, 7]. Resource-constrained workflow nets are introduced and discussed in [5, 6]. In them, the authors study extensions of workflow nets in which processes must share some global resources. A resource belongs to a type. One place is used for one type, where the resource is located when it is free. There are static and dynamic places. Static places are for resources that will be shared by cases. An important assumption in resource-constrained workflow nets is all resources are durable: they cannot be created or destroyed.

Resource-oriented Petri nets are introduced in [14, 15], which deal with finite capacity Petri nets. In a resource-oriented Petri net, a transition is enabled if and only if tokens, which represent resource, in a place won't exceed the predefined capacity of the place if the transition fires, as well as there are enough tokens in each input place.

Both resource-constrained workflow nets and resource-oriented Petri nets only deal with durable resources, which are claimed and released during the workflow execution but cannot be created and destroyed [5]. However, there are a lot of workflows where task execution consumes and/or produces resources. For example, in an incident response workflow, some resources (e.g. medication) can be consumed, and some resources (e.g. fire trucks) can be replenished during emergency response.

Besides Petri nets, some other approaches are also used for workflow resource modeling by researchers. However, their focus is on resource presentation and/or resource allocation policies. For example, a generic meta-model that can not only represent any resources in the workflow activity, but also facilitates the assigning policy implementation and execution is presented in [8]. To help improve the communication between business analysts and stakeholders, a resource behavior oriented modeling and visualization approach, based on hierarchical task network technique, for resource patterns is proposed in [4]. There are also studies that model the resource-task relationship, such as YAWL [1] and BPEL4PEOPLE [10]. But none of these works deals with workflow resource requirement analysis.

Based on our earlier work on workflow modeling [12, 13], in which a WIFA model was developed for emergency response workflows, we introduced resource oriented workflow nets (ROWN) and based on ROWN, an efficient algorithm for the analysis of the maximum resource requirement (maxRR) of a workflow was developed [11]. The maxRR is the minimum amount of resources that is needed to support workflow execution along every possible path. On the hand, when there is a resource shortage, it is important to find the minimum resource requirement (minRR), which is the minimum amount of resources that is needed to support workflow execution along at least one path. In this paper, we present an approach to analyzing the minRR.

The paper is organized as follows: Sect. 2 briefly introduces the resource-oriented workflow net model. Section 3 defines well-nested workflows and shows how it can be constructed recursively. Section 4 proposes an efficient minRR analysis algorithm. Finally, Sect. 5 concludes the paper.

2 Resource-Oriented Workflow Model

This section provides a brief introduction of ROWN. For details of ROWN, please refer to [11]. Throughout the paper, we assume that when a task gets started, all resources, if reusable, acquired by this task will be held until its execution is finished.

2.1 Task Modeling

We use two sequential transitions to model a task, one modeling the beginning of the task execution and the other the end of the task execution.

As shown in Fig. 1, transition B represents the beginning of a task execution; E represents the end of the task execution. Place $InEx$ represents the task is in execution, while $Idle$ represents the task is idle. From reachability analysis perspective, Fig. 1 can be reduced to a single transition, which represents the entire task execution as a single logic unit. This two-transition model allows the modeling of resource changes involved in task execution.

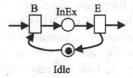

Fig. 1. Petri model of a task.

The assumption with the two-transition model is when a task gets started, the execution is guaranteed to finish. In most cases, this is true; however, there are applications in which tasks may fail in the middle of execution. For example, a machine may be broken when it is processing a job part; a data packet may get lost during transmission over a network. If a task is failed in the middle of execution, state change will be rolled back, durable resources held for the task execution will be released, but some non-durable resources may have already been consumed before its failure. In [11], we also introduced a three-transition model that models task failure. An algorithm was developed to convert a three-transition to an equivalent two-transition model in terms of resource requirement and consumption.

2.2 Resource Modeling

We assume there are n types of resources in a system and the quantity of each type of resource can be represented by a non-negative real number. Hence, resources are described by $S = (r_1, r_2, \ldots, r_n)$, where each r_i is a non-negative real number.

A task may hold or consume particular resources during execution and release or produce particular resources once execution is completed. We use R^+ to describe the resources *consumed* and/or *held* when executing a task and use R^- to describe resources *produced* and/or *released* after task execution is finished, where $r_i^+ \geq 0$ and $r_i^- \leq 0$ and $i = 1, 2, \ldots n$. Resource modeling is based on the two-transition model. More specifically, for a task TS_k, $R^+(TS_k)$ is associated with transition B_k and $R^-(TS_k)$ is associated with transition E_k. The *net consumption* of resources in executing a task is defined as

$$R^+(TS_k) + R^-(TS_k),$$

which represents the net change of available resources before and after a task is executed.

2.3   Resource-Oriented Workflow Nets (ROWN)

In [2], a Petri net that models a workflow process is called a *workflow net*. A workflow net is a Petri net that satisfies two requirements. First, it has one source place and one sink place. A token in the source place corresponds to a case needs to be handled. A token in the output place corresponds to a case that has already been handled. Secondly, there are no dangling transitions or places.

A *resource-oriented workflow net* (ROWN) is a workflow net in which the execution of a task requires resources is represented in two sequential transitions: a transition representing the start of the task and with resources occupation defined on it, and a transition representing the end of the task with resources production/release defined on it. Mathematically, it is defined as a 7-tuple: $ROWN = (P, T, R, I, O, M_0, S_0)$, where

- (P, T, I, O, M_0) is a workflow net.
- (T, R) is a pair: for each $T_k \in T$, there is an $R_k \in R$ that specifies resource change associate with the firing of transition T_k.

S_0 represents the initially available resources.

2.4   Transition Firing

A transition t_k is *enabled* under state (M_i, S_i) if and only if

$$M_i \geq I(t_k), \tag{1}$$

$$S_i \geq R(t_k) \tag{2}$$

Condition (1) stands for *control-ready*, while condition (2) stands for *resource-ready*. Notice that $R(t_k)$ is $R^+(t_k)$ if t_k represents the start of a task; it is $R^-(t_k)$ if t_k represents the end of a task. If the task represented by t_k does not involve any resource changes, then the vector is a vector of zeros.

After an enabled transition t_k fires, the new state is determined by

$$M_j = M_i + O(t_k) - I(t_k), \tag{3}$$

$$S_j = S_i - R(t_k). \tag{4}$$

Based on the transition firing rule reachability analysis can be performed, which explores all possible states and execution paths and reveals possible deadlocks due to resource contention.

2.5   An Example

Figure 2 shows an ROWN with five tasks. Assume three types of resources are involved in the workflow execution and their initial quantities are $S_0 = (30, 25, 20)$. Resource changes by task execution are specified as follows:

$$R(B_1) = R^+(TS_1) = (3, 8, 5)$$
$$R(E_1) = R^-(TS_1) = (-3, -2, -5)$$
$$R(B_2) = R^+(TS_2) = (5, 0, 10)$$
$$R(E_2) = R^-(TS_2) = (0, 0, -5)$$
$$R(B_3) = R^+(TS_3) = (15, 10, 5)$$
$$R(E_3) = R^-(TS_3) = (-5, -2, -5)$$
$$R(B_4) = R^+(TS_4) = (0, 5, 15)$$
$$R(E_4) = R^-(TS_4) = (0, -5, -5)$$
$$R(B_5) = R^+(TS_5) = (2, 0, 0)$$
$$R(E_5) = R^-(TS_5) = (-2, 0, 0)$$

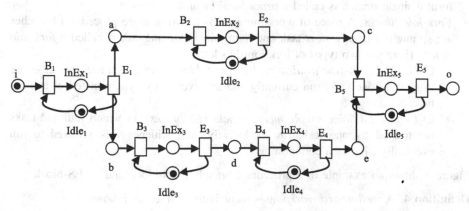

Fig. 2. An ROWN of 5 tasks.

At the initial state, B_1 is the only enabled transition. After it fires, $S_1 = S_0 - R$ $(B_1) = (27, 17, 15)$. Then E_1 is the only enabled transition. After E_1 is fired, $S_2 = S_1 - R$ $(E_1) = (30, 12, 20)$. At state (M_2, S_2), both B_2 and B_3 are enabled. If B_2 fires, $S_3 = S_2 - R$ $(B_2) = (25, 12, 10)$. Then if B_3 fires, $S_4 = S_3 - R(B_3) = (10, 2, 5)$. We can continue this process until we get to a state that no transitions are enabled, which is when place o gets a token.

3   Well-Nested Workflow Nets

3.1   Free-Choice Workflow Nets

An arbitrarily defined workflow net could be difficult to analyze because it may not possess any structural properties. In this paper we only consider *free-choice* workflow nets. A free-choice workflow net is a free-choice Petri net and formally defined as follows.

Definition 3. A workflow net is *free-choice* if and only if

$$p_1 * \cap p_2* \neq \emptyset \Rightarrow |p_1*| = |p_2*| = 1, \forall p_1, p_2 \in P,$$

Free-choice workflow nets do not allow confusion, a situation where conflict and concurrency are mixed.

3.2   Well-Nested Workflow Nets

Based on the concept of free-choice workflow nets, we further introduce *well-nested* workflow nets. To facilitate the definition of well-nested workflow nets, we introduce three building blocks:

- *Procedural Branches*. A piece of workflow in which tasks are serially connected to form a single branch is called a *procedural branch*.
- *Fork-join Blocks*. A piece of workflow in which two or more procedural branches are sprung out from a *start* task and then join in an *end* task is called a *fork-join block*. There are two types of fork-join blocks:
    ○ *Parallel Split-synchronization block* (*PS-blocks*), in which multiple tasks are trigged by one task, run currently and are eventually synchronized at another task.
    ○ *Exclusive Choice-simple merge blocks* (*ES-blocks*), in which multiple tasks are trigged by one task, run exclusively, and whichever is selected to run eventually triggers a common task.

Figure 3 shows an example of a procedural branch, a PS-block and an ES-block.

Definition 4. A *well-nested workflow* is recursively defined as follows:

1. A procedural branch is said to be a well-nested workflow.

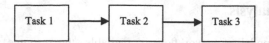

(a) A procedural branch.

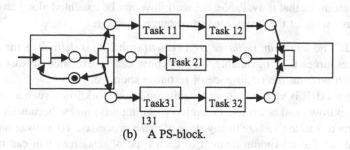

(b) A PS-block.

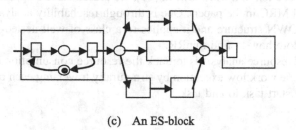

(c) An ES-block

Fig. 3. Building blocks of well-formed workflow nets.

2. A workflow resulted from replacing a task in a well-nested workflow with a fork-join block is also a well-nested workflow.

Figure 4 shows a well-nested workflow.

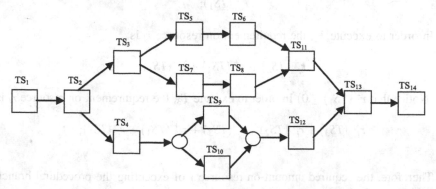

Fig. 4. A well-nested workflow net. Tasks 2–13 constitutes a PS-block with TS$_2$ and TS$_{13}$ being the start and end tasks of the block, respectively. The two branches of this PS-block are a smaller PS-block, composed of TS$_3$, TS$_{5-8}$ and TS$_{11}$ and an ES-block, composed of TS$_4$, TS$_9$, TS$_{10}$ and TS$_{12}$.

4 Resource Analysis

With the ROWN model of a workflow is available, the resource requirement for executing the workflow can be formally analyzed. We are interested in the *maximum resource requirement* (maxRR) and *minimum resource requirement* (minRR).

Definition 1. The *maximum resource requirement* (maxRR) is defined as the minimum amount of resources that if available, the workflow can be executed along any possible path till finish without the occurrence of resource shortage.

Definition 2. The *minimum resource requirement* (minRR) is defined as the minimum amount of resources that if available, the workflow can be executed at least along one path till finish without the occurrence of resource shortage.

Meeting maxRR is very important for mission critic workflows, such as emergency response workflows and enterprise exception handling workflows, because it is desired to not see any resource shortage in any stage of these processes. To analyze maxRR, we need to find out the maximum amount of each type of resources that can be held or consumed in the execution of a workflow. In [11], we presented two approaches to find maxRR (it is called MRC in the paper). One is through reachability analysis; the other one is based on ROWN structure and it applies to a class of well-nested ROWNs. In this paper, we explore how to find minRR.

The purpose of resource analysis is to track the resource consumption of each type of resource during the workflow execution by structurally traversing each branch of the workflow from the start task to end task.

4.1 Procedural Branches

Analysis of minRR for a well-nested workflow net is mainly based on the analysis of procedural branches. Consider a procedural branch constituting of tasks TS_1, TS_2, ..., TS_p, sequentially. In order to execute TS_1, the requirement on resource i is:

$$r_i^+ (TS_1);$$

In order to execute T_2, the requirement on resource r_i is:

$$r_i^+ (TS_2) + r_i^+ (TS_1) + r_i^- (TS_1);$$

(Notice that $r_i^- (TS_1) \leq 0$) In order to execute T_3, the requirement on resource r_i is:

$$r_i^+ (TS_3) + r_i^+ (TS_2) + r_i^- (TS_2) + r_i^+ (TS_1) + r_i^- (TS_1);$$

$$\cdots$$

Therefore, the required amount on resource i of executing the procedural branch, denoted by $Rq(i)$, is

$$Rq(i) = \max\{r_i^+(TS_1),$$
$$r_i^+(TS_2) + r_i^+(TS_1) + r_i^-(TS_1),$$
$$\cdots$$
$$r_i^+(TS_p) + \sum_{k=1}^{p-1}\left(r_i^+(TS_k) + r_i^-(TS_k)\right)\} \tag{5}$$

The net resource consumption is

$$Rc(i) = \sum_{k=1}^{s}\left(r_i^+(TS_k) + r_i^-(TS_k)\right) \tag{6}$$

The overall required resources of executing the procedural branch is

$$R_r = (Rq(1), Rq(2), \ldots Rq(n)) \tag{7}$$

4.2 ES-Blocks

For an ES-block, the idea is to find a branch that requires the least amount of resources.

Consider an ES-block composed of w procedural branches, we use Eq. (7) to calculate the resource requirement for each branch, excluding the start and end tasks. Denote by $Rq_k(i)$ the requirement on resource i of branch k. Our goal is to find a branch that takes

$$\min(Rq(1), Rq(2), \ldots Rq(s))$$

among all branches. If

$$Rq_k(i) = \min\{Rq_j(i)|j = 1, 2, \ldots, w\} \text{ for } i = 1, 2, \ldots, n, \tag{8}$$

Then branch k consumes the least amount of resources of every type among all branches.

However, it is not always possible to find a branch in an ES-block that satisfies the condition described by (8). It is rather normal that one branch uses more resources of one type while another branch uses more resources of another type. One way we can use to handle this situation is prioritize different types of resources according to some criteria, such as availability, accessibility, cost, or a mix of them. Then we select a branch that would use the least amount of resources with the highest priorities.

A more general way of implementing this idea is take a linear scalarization of resources required. We are looking for a branch that gives

$$\min \sum_{i=1}^{n} w_i Rq(i) \tag{9}$$

where the weights of the resources $w_i \geq 0$ are parameters of the scalarization.

When the branch, say branch k, that requires the minimum resources is found, then we can calculate its resource consumption using Eq. (6).

4.3 PS-Blocks

For a PS-block, the idea is to find a path along which all tasks in the block are executed and the least amount of resources is required.

We attempt to find such a path though reachability analysis. Consider a PS-block that is composed of w procedural branches. Since these branches are in parallel, all of them have to be executed. Assume there are v tasks in total on those w branches, excepting the start and end tasks of the block. Let H be the reachable state set of the PS-block with the start and end tasks removed, which results in a ROWN of all these parallel and disjoint branches. Generate a reachability tree for this ROWN. Each path from the root to a leaf represents an execution path, which constitutes all $2v$ transitions in the block (Recall each task is modelled with two transitions). Notice that in any given path tasks from different branches may be interwoven. Moreover, it represents true concurrency: before one task is finished, other tasks may start execution.

Without loss of generality, we denote an execution path by a sequence of transitions $t_1 t_2 \ldots t_{2v}$. To fire transition t_1, the resource required for type i is

$$r_i(t_1);$$

After that, to fire t_2, the resource required is

$$r_i(t_1) + r_i(t_2);$$

To further fire t_3, the resource required is

$$r_i(t_1) + r_i(t_2) + r_i(t_3);$$
$$\ldots$$

Therefore, the required amount on resource i of executing the path, denoted by $Rq(i)$, is

$$
\begin{aligned}
Rq(i) = \max\{ & r_i(t_1), \\
& r_i(t_1) + r_i(t_2), \\
& r_i(t_1) + r_i(t_2) + r_i(t_3); \\
& \ldots \\
& \sum_{k=1}^{2v} r_i(t_k) \}
\end{aligned}
\tag{10}
$$

The overall required resources of executing the path is

$$R_r = (Rq(1), Rq(2), \ldots Rq(n)) \tag{11}$$

The net resource consumption is

$$Rc(i) = \sum_{(k=1)}^{2v} r_i(t_k) \tag{12}$$

Assume the reachability tree indicates that there are q execution paths. We can treat these q execution paths as an ES-block that has q branches, each branch being an execution path. Then we apply the strategy for ES-block analysis to find the path that requires the minimum amount resources.

Example. Consider again the workflow shown in Fig. 2. It is a PS-block, with TS_1 and TS_5 being the start and end tasks, respectively. The PS-block has two procedural branches: the first branch has only task 2 and the second branch has task 3 and task 4.

According to Eqs. (7) and (9), the resource requirement of the first procedural branch is $R(B_1) = R^+(TS_1) = (3, 8, 5)$ and that of the second procedural branch is $R(B_5) = R^+(TS_5) = (2, 0, 0)$. For the PS-block, there are 15 possible execution paths and they are listed as follows, where paths 1–5 are obtained by firing B_2 first and varying the firing order of E_2, paths 6–9 are obtained by firing B_3B_2 first and varying the firing order of E_2, paths 10–12 are obtained by firing $B_3E_3B_2$ first and varying the firing order of E_2, paths 13-14 are obtained by firing $B_3E_3B_4B_2$ first and varying the firing order of E_2, and paths 15 is obtained by firing $B_3E_3B_4E_4B_2$ first.

$$Path\ 1 : B_2E_2B_3E_3B_4E_4$$
$$Path\ 2 : B_2B_3E_2E_3B_4E_4$$
$$Path\ 3 : B_2B_3E_3E_2B_4E_4$$
$$Path\ 4 : B_2B_3E_3B_4E_2E_4$$
$$Path\ 5 : B_2B_3E_3B_4E_4E_2$$
$$Path\ 6 : B_3B_2E_2E_3B_4E_4$$
$$Path\ 7 : B_3B_2E_3E_2B_4E_4$$
$$Path\ 8 : B_3B_2E_3B_4E_2E_4$$
$$Path\ 9 : B_3B_2E_3B_4E_4E_2$$
$$Path\ 10 : B_3E_3B_2E_2B_4E_4$$
$$Path\ 11 : B_3E_3B_2B_4E_2E_4$$
$$Path\ 12 : B_3E_3B_2B_4E_4E_2$$
$$Path\ 13 : B_3E_3B_4B_2E_2E_4$$
$$Path\ 14 : B_3E_3B_4B_2E_4E_2$$
$$Path\ 15 : B_3E_3B_4E_4B_2E_2$$

We apply Eq. (12) to calculate the resource requirement of path 1 as follows:

$Rq(1) = \max\{5, 5+0, 5+0+15, 5+0+15+(-5), 5+0+15+(-5)+0, 5+0+15+(-5)+0+0\} = 20;$
$Rq(2) = \max\{0, 0+0, 0+0+10, 0+0+10+(-2), 0+0+10+(-2)+5, 0+0+10+(-2)+5+(-5)\} = 13;$
$Rq(3) = \max\{10, 10+(-5), 10+(-5)+5, 10+(-5)+5+(-5), 10+(-5)+5+(-5)+15,$
$\qquad 10+(-5)+5+(-5)+15+(-5)\} = 20;$

Therefore, the resource requirement of path 1 is (20, 13, 20). Using the same approach, we can find the resource requirements of other paths are as follows:

$$\text{Paths } 2, 3, 6, 7, 10, 15 : (20, 13, 20)$$
$$\text{Paths } 4, 5, 8, 9 : (20, 13, 25)$$
$$\text{Paths } 11, 12, 13, 14 : (15, 13, 25)$$

It is obvious that executing the PS-block along paths 4, 5, 8 or 9 requires the largest amount of resources. On the other hand, path 1, 2, 3, 6, 7, 10, or 15 requires more resource of type 1 but less resource of type 3 than path 11, 12, 13, or 14. In this case, we can select a path between these two groups based on the availability or any other criteria on the two types of resources. For example, if there is abundant resource of type 1 available meanwhile it is harder to supply the resource of type 3, then we can select any path in the first group for the PS-block execution, because it requires less resource of type 3. Based on Eq. (14), the net resource consumption of executing the block is

$$(5, 010) + (0, 0, -5) + (15, 10, 5) + (-5, -2, -5) + (0, 5, 15) + (0, -5, -5)$$
$$= (15, 8, 15).$$

4.4   Algorithm for minRR Analysis

The overall workflow minRR analysis is carried out by the analysis of PS-blocks and ES-blocks and then replacing all branches of each block with a task that is equivalent in terms of minimum resource requirement and resource consumption. It starts from internal-most blocks. Before we formally describe the algorithm, let us illustrate the idea using the workflow of Fig. 4 first. In this workflow, there are two internal most blocks:

- ES-block, composed of TS_4, TS_9 and TS_{10} and TS_{12}.
- PS-block, composed of TS_3, TS_5, TS_6, TS_7, TS_8 and TS_{11}.

For the ES block, we replace the two branches with a new task T_{102}, while for the PS-block, we replace the two branches with a new task T_{101}, as illustrated in Fig. 5. Now there is only one block in Fig. 5, which starts with TS_2 and ends with TS_{13} and has two branches, each having three tasks. We replace each branch with a single new task, and then replace the two branches (now each branch has only one task) with a new task. After that, only one branch is left in the ROWN.

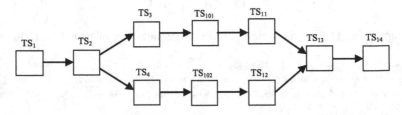

Fig. 5. An equivalent ROWN to Fig. 4.

For any type of blocks, when all its branches are replaced with a single task TS_b, the equivalent resource parameters of TS_b for resource i are:

$$r_i^+ (TS_b) = Rq(i) \tag{13}$$

$$r_i^- (TS_b) = Rq(i) - R_c(i) \tag{14}$$

If the block is an ES-block, $Rq(i)$ and $R_c(i)$ are calculated based on the approach described in Sect. 4.2. If it is a PS-block, they are calculated using the approach described in Sect. 4.3.

minRR analysis algorithm

Input: A resource oriented workflow

Output: The minimum requirement on any type of resource i to allow the workflow execute along at least one path.

Step 1: Identify all blocks.

Step 2: If no blocks exist, go to Step 4.

Step 3: Identify an internal-most block.

Step 3.1: Replace all branches in the block with a single task T_b, that is connected to the start and end tasks of the block, calculate its $r_i^+(TS_b)$ and $r_i^-(TS_b)$ for all resource type i using Eqs. (13) and (14).

Step 3.2: Go to *Step 1*.

Step 4: Now the workflow becomes a procedural branch. Use Eq. (7) to calculate R_r. Output R_r.

The minRR analysis is scalable: It can handle large size of workflow nets efficiently by repeating the aggregation process. In each round a set of tasks is replaced with a single task, so the workflow size shrinks rapidly.

5 Concluding Remarks

Petri nets are a powerful tool in modeling and analyzing resources-constrained systems. Workflow nets are a type of Petri nets. Resource-oriented workflow nets (ROWNs) are workflow nets incorporated with resource definition. Based on ROWN model, this paper discussed the analysis of minimum resource requirement (minRR) for workflow execution. An efficient algorithm for minRR analysis is presented for well-nested workflows. This algorithm works by identifying blocks in a workflow and replacing an internal-most block with a single task a time, where the task's resource parameters are calculated based on approaches developed in this paper to ensure the new workflow is equivalent to the old one in terms of minRR. The process repeats until the workflow becomes a procedural branch and then the minRR can be calculated using a simple formula.

Acknowledgement. The work presented in this paper is partially supported by Mexico CONACYT program "Estancia Sabatica en Mexico para Eranjeroa."

References

1. van der Aalst, W.M.P., ter Hofstede, A.H.M.: YAWL: yet another workflow language. Inf. Syst. **30**(4), 245–275 (2005)
2. van der Aalst, W.M.P.: Verification of workflow nets. In: Azéma, P., Balbo, G. (eds.) ICATPN 1997. LNCS, vol. 1248, pp. 407–426. Springer, Heidelberg (1997)
3. Fanti, M.P., Zhou, M.: Deadlock control methods in automeated manufacturing systems. IEEE Trans. Syst. Man Cybern. Part A **34**(1), 5–21 (2004)
4. Guo, H., Brown, R., Rasmussen, R.: Workflow resource pattern modelling and visualization. In: Australasian Computer Science Week, 29 January – 1 February, Adelaide, Australia (2013)
5. van Hee, K., Sidorova, N., Voorhoeve, M.: Resource-constrained workflow nets. Fundamenta Informaticae **71**(2–3), 243–257 (2005)
6. Juhás, G., Kazlov, I., Juhásová, A.: Instance deadlock: a mystery behind frozen programs. In: Lilius, J., Penczek, W. (eds.) PETRI NETS 2010. LNCS, vol. 6128, pp. 1–17. Springer, Heidelberg (2010)
7. Martos-Salgado, M., Rosa-Velardo, F.: Dynamic soundness in resource-constrained workflow nets. In: Bruni, R., Dingel, J. (eds.) FORTE 2011 and FMOODS 2011. LNCS, vol. 6722, pp. 259–273. Springer, Heidelberg (2011)
8. Zur Muehlem, M.: Resource modeling in workflow applications. In: Workflow Management Conference (1999)
9. Russell, N., van der Aalst, W.M., ter Hofstede, A.H., Edmond, D.: Workflow resource patterns: identification, representation and tool support. In: Pastor, Ó., Falco e Cunha, J. (eds.) CAiSE 2005. LNCS, vol. 3520, pp. 216–232. Springer, Heidelberg (2005)
10. Russell, N., van der Aalst, W.M.P.: Work distribution and resource management in BPEL4People: capabilities and opportunities. In: Bellahsène, Z., Léonard, M. (eds.) CAiSE 2008. LNCS, vol. 5074, pp. 94–108. Springer, Heidelberg (2008)
11. Wang, J., Li, D.: Resource oriented workflow nets and workflow resource requirement analysis. Int. J. Softw. Eng. Knowl. Eng. **23**(5), 667–693 (2013)
12. Wang, J., Tepfenhart, W., Rosca, D.: Emergency response workflow resource requirements modeling and analysis. IEEE Trans. Syst. Man Cybern. Part C **39**(3), 270–283 (2009)
13. Wang, J., Rosca, D., Tepfenhart, W., Milewski, A., Stoute, M.: Dynamic workflow modeling and analysis in incident command systems. IEEE Trans. Syst. Man Cybern. Part A **38**(5), 1041–1055 (2008)
14. Wu, N., Zhou, M.: Resource-oriented petri nets in deadlock avoidance of AGV systems. In: Proceedings of the 2001 IEEE International Conference on Robotics and Automation, Seoul, Korea, 21–26 May 2001
15. Wu, N., Zhou, M.: System Modeling and Control with Resource-Oriented Petri Nets. CRC Press, Boca Raton (2009)

Comparison and Variability Analysis in Process Variants

Jimin Ling$^{(\boxtimes)}$ and Li Zhang

School of Computer Science and Engineering,
Beihang University, Beijing, China
lingjimin@buaa.edu.cn

Abstract. The co-existence of multiple variants of a same process is a common phenomenon in process repositories of organizations. To manage these process variants effectively, it is necessary to compare them against each other and identify the commonalities and differences between them. However, existing process matching techniques are mostly limited to a comparison between two process variants only. Comparing a set of process variants to identify common and variant elements becomes thus a major challenge. In this paper, we propose the Process Variants Comparison (ProVC) method to compare and analyze variability between a set of process variants rather than binary comparison of two variants. An intuitive and intelligible visualization of the comparison result is then provided to users in the form of presence and relative position of nodes in process variants. The accuracy and usability of the ProVC method are demonstrated by a case study and user experiment in an application scenario.

Keywords: Process model · Model comparison · Process variants · Variability · Process match

1 Introduction

Process models are becoming increasingly widespread in organizations as a result of the more and more attention paid to their processes. The co-existence of multiple variants of a same process is a common phenomenon in process model repository of these organizations. This is probably caused by different legal requirements in different areas, deviations in the IT infrastructure, or organizational differences [1].

As a concrete example, *Capital Airport VIP Service Management Co., Ltd.* (*VIP Company* for short), the largest airport VIP service company in China, provides different service portfolio for different kinds of customers, such as national leaders, government VIPs, commercial VIPs, commercial CIPs, and other individual members. These services exist in different subsidiary companies (i.e. airport management companies in other Chinese provinces) and they are generally similar but slightly different from each other. As a result, there are tens of process variants for VIP service in the company.

With the background of co-existence of process variants, it is meaningful to compare the existing process variants from each other and identify the commonalities and differences between them. Carrying out such work is a foundation of keeping these

© Springer Science+Business Media Singapore 2016
J. Cao et al. (Eds.): PAS 2015, CCIS 602, pp. 67–78, 2016.
DOI: 10.1007/978-981-10-1019-4_6

process variants co-evolved [2] and a prerequisite for any process harmonization effort that aims at reducing the amount of allowed process variability [3].

Doing above comparison work manually is time-consuming, thus some automatic matching techniques [4–9] are proposed to construct correspondences between process variants based on activity label semantics and model structure. These methods could effectively handle the challenge of lexical and semantic diversity on element labels and different model abstraction levels when identify the common and different parts in process variants. However, these methods are limited to a comparison between two process variants only. Comparing all models among a set of process variants to identify common and variant elements could help business analyzer to get a whole picture of these variants rather than comparing two models one by one. Besides, the intuitive and convenient visualization of the process variant comparison result is neglected in the existing approaches.

To deal with the challenges mentioned above, this paper proposes a new method, which is called *Process Variant Comparison* (*ProVC*), to identify the commonalities and variabilities between a set of process variants rather than compare them one by one. Besides, in order to provide an intuitive visualization of the variants comparing result to users, we provide a visual extension of process modeling notations to present the presence and relative position of nodes in these process variants. Finally, a case study and a user experiment based on process variants of *VIP Company* are conducted to evaluate the accuracy and usability of the *ProVC* approach.

The rest of paper is organized as follows: The next section gives a running example used throughout the paper to illustrate our approach. Section 3 presents our approach in detail. Section 4 reports the case study of the approach. Section 5 presents the related work and Sect. 6 concludes this paper and presents the future work.

2   Running Example

As a concrete example, we use throughout this paper a set of EPC process model variants which represent a set of booking and service process in *VIP Company*. The running example is shown in Fig. 1. Four simplified models of these process variants are given here due to the limited space. The booking and service procedure starts with a booking request from service phone. For national leaders and some government VIPs, the booking request may come from a special phone line (*variant c*). The customer's identity is then verified by self-service or manual-voice service. Then the order request will be submit to different booking center according to different types, e.g. booking request for a government VIP cannot be submitted to junior booking center (*variant c*), while another possible condition is that some subsidiary companies only have one booking center (*variant b*). After the user submits the information of flight itinerary, time-limit (e.g. booking should be at least two hours before the flight departure and the flight date should not exceed the membership period) will be checked in some circumstances (*variant a* and *d*). The expense settlement is handled before (*variant a*) or after (*variant b, c* and *d*) the customer received the VIP service. When the customer

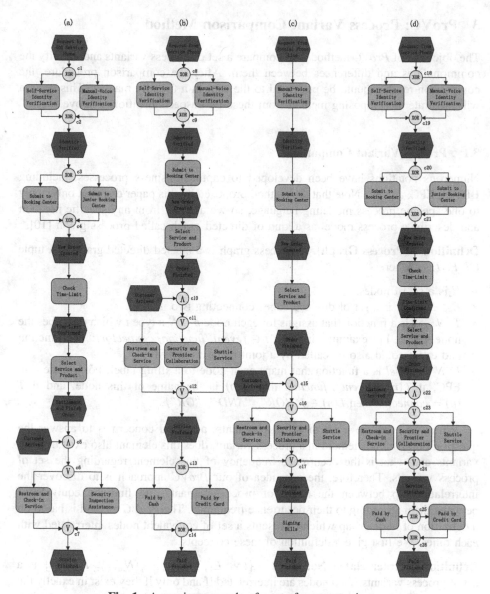

Fig. 1. A running example of a set of process variants

arrives at the airport, the services always include some basic items such as restroom and check-in services and assist of security inspection, while some other services such as shuttle service and frontier inspection collaboration are optional (*variant b, c* and *d*). Finally, after all these services are done, each process variant provides different terms of payment including cash and credit card (*variant b, c* and *d*). For the government VIPs, only the manner of signing bills is required (*variant c*).

3   ProVC: Process Variant Comparison Method

The objective of *ProVC* method is to compare a set of process variants and identify the commonalities and differences between them. After the comparison procedure, the comparison results should be presented to the user in a friendly manner. This section will illustrate the proposing method from these two aspects mentioned above.

3.1   Process Variant Comparison

Numerous notations have been developed to capture business processes, including BPMN, EPC, and etc. Note that the method proposed by this paper does not only apply to one specific process modeling language, so we abstract from any specific notation and describe a process model as a kind of directed graph called process graph [10].

Definition 1 (Process Graph). A process graph is a labeled directed graph in a tuple (N, E, T, L) where:

- N is a set of nodes;
- $E \in N \times N$ is a set of directed edges connecting two nodes;
- $T: N \rightarrow t$ is a function that assigns for each node $n \in N$ a type t which describes the node's type. For example, in EPC, $t \in \{event, function, connector\}$; The function and event could also be called by a joint name as activity;
- $L: N \rightarrow label$ is a function that maps each node to a string label; For a node n in EPC, if $T(n) = event \vee function$ then $L(n)$ is the name of this node, and if $T(n) = connector$ then $L(n) \in \{"XOR", "AND", "OR"\}$.

When we compare a set of process variants, a critical concern is to answer the question of, given an element in a process variant, does this element also exist in other variants and what is the occurrence frequency of this element regarding the set of process variants. Therefore, the core idea of our *ProVC* approach is to discover the interrelationship between the nodes of process variants and find the equivalence between them according to their occurrence frequency. The *ProVC* method is based on a definition of node group which represents a set of equivalent nodes interrelated with each other. We first give a definition of these concepts.

Definition 2 (Interrelated Nodes). $S = \{(N_1, E_1, T_1, L_1), \ldots, (N_m, E_m, T_m, L_m)\}$ is a set of process variants. Two nodes are interrelated if and only if they exist in exactly the same process variants. In other words, two nodes n_i and n_j in process variant (N_i, E_i, T_i, L_i) and (N_j, E_j, T_j, L_j), $n_i \in N_i$, $n_j \in N_j$ then n_i and n_j are interrelated, denoted as $n_i \sim n_j$, if they satisfy the following condition:

$$\forall (N_k, E_k, T_k, L_k) \in S \mid n_i \prec N_k \Leftrightarrow n_j \prec N_k$$

In the above formula, $n \prec N$ means that node n exists in the node set N of the process variant (N, E, T, L). Because one same activity could have different but similar labels in different variants, e.g. the event *"Request by 400 Service Phone"* in *variant a* and event *"Request from Service Phone"* in *variant b* in Fig. 1, we need to consider node similarity when judging a specific node exists in a process variant or not. Given one specific node, we consider this node exists in a process variant if there is a node in the process variant which is identical or similar to the given node.

Definition 3 (Node Exist Relationship). Given a node n and a process variant (N, E, T, L), n is said to be exist in process variant (N, E, T, L), which is denoted as $n \prec N$, if they satisfy the following condition:

$$\exists n' \in N \mid SimN(n, n') \geq cutoff$$

In the above formula, $SimN(n, n')$ means the node similarity value of n and n'. How to measure the similarity between two specific nodes will be discussed later in next section. When the similarity value achieves a given *cutoff* value, we consider these two nodes are similar. We set this *cutoff* value to 0.75 according to the experiment results in our previous work [6].

We can see that the interrelated relationship between nodes could divide all the nodes in process variants into several parts according to their occurrence frequency in the set process variants, so we define the notion of node group as follows.

Definition 4 (Node Group). $S = \{(N_1, E_1, T_1, L_1), \ldots, (N_m, E_m, T_m, L_m)\}$ is a set of process variants. A node group of S is the maximum set of interrelated nodes, denoted as $NG(S)$, i.e. satisfy both of the following conditions:

$$\forall n_i, n_j \in NG(S) \mid n_i \sim n_j$$

$$\forall n_i \notin NG(S), \forall n_j \in NG(S) \mid \neg(n_i \sim n_j)$$

Here we use the running example to explain the concepts of interrelated nodes and node group. We could easily find that the event *"New Order Created"* and function *"Select Service and Product"* exist in all of the four process variants, so these two nodes are interrelated. Besides, the event *"Request by 400 Service Phone"* (including the corresponding event *"Request from Service Phone"* in *variant b*) and function *"Self-Service Identity Verification"* are both appeared in *variant a, b*, and *d*, thus they are interrelated but belong to another node group. The identification of all node groups of the process variants are presented in Fig. 2. The *Node Group 1* gathers all the nodes that are present in all the process variants. The *Node Group 2* and *3* concern the differences of booking and payment between government VIPs and other customers. The implications of other node groups could also be analyzed from Fig. 2.

3.2   Node Similarity Measurement

Here we resolve the remaining issue of node similarity measurement which is mentioned above. This issue could be described as follows: given two nodes from two process graph respectively, how to calculate the value (the range of [0,1]) representing their similarity degree. In this paper, we only consider the potential similarity of two

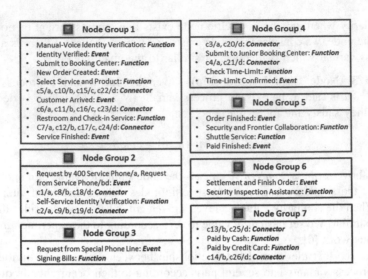

Fig. 2. Node groups of the process variants

nodes in case they have same type. Based on this assumption, we need to compute the similarity between activities (*SimA*) and the similarity between connectors (*SimC*).

For similarity between activities, we use a combination of two lexical metrics which are widely used in process matching research area, i.e. syntactic and semantic similarity. The syntactic similarity based on *Levenshtein Distance* which is the number of atomic string operations (i.e. insert, delete or substitute a character) necessary to get from one string to another. The semantic similarity is based on equivalence between the words they consist of, i.e. taking synonyms into consideration. The detail computing formula of these two concepts could be found in [11]. We consider two activities are similar if they are either syntactic or semantic similar, i.e. *SimA* could be computed as the higher value of the syntactic and semantic similarity between them.

For similarity between two connectors, it is unrealistic to measure them by lexical similarity because they do not have any textual information. Instead, we use a notion of context similarity which means the matching degree of their presets and postsets. More detail of connector similarity *SimC* could be found in our previous work [6, 7].

3.3   Visualization of Comparison Results

After all the comparison works of process variants have done, it is important to present an intuitive visualization of the comparison results while this aspect is usually neglected by most existing process matching studies. Here we propose a visual extension of the basic process modeling notation to visualize presence and relative position of nodes in these process variants.

Figure 3 shows a part of comparison result of our running example. Only two variant are presented here due to the limitation of space. This representation is based on extension of original EPC model. Nodes in process variants are marked by different background color according to the node groups they belonging to. Users could easily

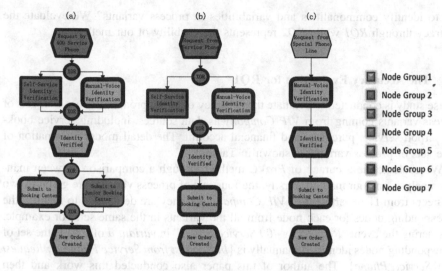

Fig. 3. A part of comparison result of running example (Color figure online)

identify the same class of nodes and their interrelations by the background colors. Besides, the boundary lines of these nodes have different linestyles based on the frequency of existence in all process variants. For example, the nodes in *Node Group 1* have the thickest linestyle because they exist in all process variant. These above visual extension of process variants provide an intuitive representation to users to make them understand the comparison results more easily.

We also provide a table which represents the relation of process variants and node groups for further visualization and understanding. This relation is presented in Table 1.

Table 1. Relations of process variants and node groups

	Variant a	Variant b	Variant c	Variant d
Node group 1	√	√	√	√
Node group 2	√	√		√
Node group 3			√	
Node group 4	√			√
Node group 5		√	√	√
Node group 6	√			
Node group 7		√		√

4 Evaluation

In this section, we will evaluate our *ProVC* method from two research questions: *RQ1*: Is the *ProVC* method able to produce correct comparison and variability analysis results? *RQ2*: Does *ProVC* method indeed provide convenience for users when they aim to identify commonalities and variabilities in process variants? We evaluate the accuracy through *RQ1* while *RQ2* represents the usability of our method.

4.1   The Accuracy Evaluation for RQ1

A case study is conducted to evaluate the accuracy of our approach. We use four sets of process variants coming from *VIP Company* as data sources, including service booking, airport service, purchase, and financial account. The detail model information of these sets of process variants is shown in Table 2.

We evaluate the accuracy of *ProVC* method through a comparison between manually analysis and our method. Firstly, the four sets of process variants are given to two engineers from IT department of *VIP Company* and they are demanded to identify the corresponding nodes for each node from all the variants in the same set. For example, considering the event "*Request by 400 Service Phone*" in *variant a* of Fig. 1, the set of corresponding nodes identified manually is {*b: Request from Service Phone, d: Request from Service Phone*}. The author of this paper also conducted this work and then discussed with the two engineers to reach an agreement about the comparison result. Comparing this manual decided result with the results obtained by *ProVC* method, every node in the set of variants is judged one by one to see whether these two results for this node are matched. The comparison results for each set are shown in Table 3. The evaluation shows that the *ProVC* method could produce a good model comparison result in this industry case.

Table 2. Statistics of process variants for case study

ID	Variants sets	Variants	Avg. nodes	Min. nodes	Max. nodes
1	Booking	6	25.3	18	32
2	Service	5	16.0	13	20
3	Purchase	4	13.5	11	17
4	Finance	5	17.0	11	25

Through further study to the miss-matched results between *ProVC* method and manual judgment, we found that the fault comparison results mainly come from two circumstances. The first one is that different words and phrases are used to describe the same activity. This is a common challenge of all process matching researches and we will apply some other novel node similarity metrics to improve this situation in the future. Other few negative results are caused by the similar activity labels with distinct meaning. For example, considering the function "*Submit to Junior Booking Center*" in *variant a* of Fig. 1, it should be a common part of *variant a* and *d*, but our method matched it with "*Submit to Booking Center*" incorrectly and judge it as a common part of all the four variants.

Table 3. Accuracy of *ProVC* method in case study

ID	Node groups num.	Right nodes num.	Fault nodes num.	Accuracy rate
1	11	139	13	91.4 %
2	7	73	7	91.25 %
3	8	54	0	100 %
4	8	80	5	94.1 %

4.2 The Usability Evaluation for RQ2

A user experiment is conducted to evaluate the usability of our approach. The subjects are 20 graduated students from School of Software in Beihang University. All of them have at least two years of software modeling with UML. They are randomly divided into *Group A* and *B* in average. The data sources of this experiment are still the four sets of process variants from *VIP Company*. Ten questions related to model comparison are designed for each set. Sample questions include "Does *function A* in *variant a* and *function B* in *variant b* express the same business logic?" "Which variants need to perform the task of checking time-limit of member identity?" "Which activities are only required to be performed for government VIPs?" "Which functions are contained in all process variants in this set?" All subjects are demanded to answer all these forty questions with the assist of *ProVC* model comparison results (called *Assist P*) or conventional process matching results (called *Assist Q*) provided by [4].

Before the subjects conduct the experiment, we give them a brief introduction about process modeling and EPC modeling language. Then the process comparison tasks are expounded to them. After these preparations, two steps are executed. Firstly, subjects from *Group A* are demanded to deal with questions about *Variant Set 1* and *2* using *Assist P* and meanwhile subjects from *Group B* deal with the same questions using *Assist Q*. After all subjects finished the questions in the first step, subjects from Group A are demanded to deal with questions about *Variant Set 3* and *4* using *Assist Q* and subjects from *Group B* deal with the same questions using *Assist P* at the same time. All subjects are demanded to record the time when they finish questions of each variant set. We compute the average values of correctness and time-using with different assisting methods and the p-values comparing them. The results are shown in Table 4.

Table 4. Usability evaluation of *ProVC* method

Variants sets		Correctness rate		Time-using (min)	
		Average	P-value	Average	P-value
Variants *set 1*	Assist P	0.95	0.2475	14.1	1.30E-4
	Assist Q	0.91		21.8	
Variants *set 2*	Assist P	0.97	0.739	11.0	2.06E-4
	Assist Q	0.98		13.9	
Variants *set 3*	Assist P	0.98	1.0	10.6	3.90E-3
	Assist Q	0.98		13.3	
Variants *set 4*	Assist P	0.96	0.4813	10.9	4.87E-4
	Assist Q	0.94		13.9	

From the results we can see that the subjects could answer the questions with high correctness rate by using both of the assisting method, but the significant short time-using is achieved by adopting *ProVC* method to assist the model compassion and variability analysis procedure, thus the usability of *ProVC* method is demonstrated to a certain degree.

5 Related Work

Various research works [4–9] have tried to compare and match process models based on automated techniques. *Dijkman et al.* [4] compare three process matching approaches based on lexical or graph matching: activity label matching, greedy graph matching and A* graph matching. The result shows that the greedy graph matching technique produces the best precision while the activity label matching has advantage on efficiency. The node similarity measurement used in this paper is similar to activity label matching in [4] which is a simple and efficient way to match model nodes, and how to improve precision and recall of the building matches is out of scope of this paper. The ICoP framework [5] enables the optional creation of process matchers from the reusable components to provide a strong extendibility, but the implementation of the components have not been illustrated clearly. Our earlier researches [6, 7] propose a process matching technique to support fast detecting node correspondences based on extended graph edit distance and tree edit distance. *Klinkmuller et al.* [8] propose an optimal activity label matching method utilizing the idea of Bag-of-Words and Label-Pruning to get higher precision than the traditional syntactic and semantic similarity metrics. *Branco et al.* [9] propose a process matching approach which aims to construct correspondences between model fragments rather than nodes. All these related researches mentioned above concentrate on increasing accuracy of node matching algorithm or supporting construction of complex matching relations (e.g. matching process fragments), but how to identify variabilities in a set of process variants instead of a pair of models and how to present the comparing results more intuitively are not clearly considered yet.

Beyond the process modeling domain, process variants comparison is also related to the research area of software model compare. EMF Compare [12] has been proposed based on Eclipse framework to compare software models. It integrates several statistics and metrics to match the model elements and the comparison result is visualized by highlights model fragments. *Rubin and Chechik* [13] study how to compare and merge UML model variants into a consolidated one, which is similar to process model merge [10], but they do not provide visualization of comparison results. Besides, some comparison approaches [14–16] are proposed for models that can conform to an arbitrary meta-model, assuming it adheres to specific properties. Most of these above studies are lack of the ability to handle the situation of the co-existence of a set of model variants, i.e. compare all model variants together at the same time. In contrast, our *ProVC* method not only supports the comparison and variability analysis between a set of process variants at the same time, but also provides an intuitive and intelligible representation of the comparison result.

6 Conclusions

We proposed a process variant comparison method to identify the commonalities and variabilities between a set of process variants rather than binary comparison of two variants. An intuitive and intelligible visualization of the comparing result is then provided to users in the form of presence and relative position of nodes in process variants. We validated our approach base on a case study and a user experiment to demonstrate the accuracy and usability of the *ProVC* method. The evaluation results show that the *ProVC* approach could produce good model variability analysis result with high accuracy and provide significant convenience for users when they aim to identify commonalities and variabilities in process variants.

In the future, we aim to apply some novel node similarity metrics to our *ProVC* methods to improve the accuracy of comparison result. Besides, we will apply *ProVC* method to large scale of process variants in other industrial scenario to further evaluate the usability of our approach.

Acknowledgment. This work is supported by the National Natural Science Foundation of China (No. 61170087 and No. 61370058). Thanks for all participants involved in the evaluation of this work, including graduated students from School of Software in Beihang University and the engineers from IT department of the Capital Airport VIP Service Management Co., Ltd.

References

1. Wijnhoven, F., Spil, T., Stegwee, R., et al.: Post-merger IT integration strategies: an IT alignment perspective. J. Strateg. Inf. Syst. **15**(1), 5–28 (2006)
2. Weidlich, M., Mendling, J., Weske, M.: Propagating changes between aligned process models. J. Syst. Softw. **85**(8), 1885–1898 (2012)
3. Weidlich, M., Mendling, J., Weske, M.: A foundational approach for managing process variability. In: Mouratidis, H., Rolland, C. (eds.) CAiSE 2011. LNCS, vol. 6741, pp. 267–282. Springer, Heidelberg (2011)
4. Dijkman, R., Dumas, M., Garcia, B.L., Kaarik, R.: Aligning business process models. In: Proceedings of the 13th IEEE International Enterprise Distributed Object Computing Conference, pp. 45–53 (2009)
5. Weidlich, M., Dijkman, R., Mendling, J.: The ICoP framework: identification of correspondences between process models. In: Pernici, B. (ed.) CAiSE 2010. LNCS, vol. 6051, pp. 483–498. Springer, Heidelberg (2010)
6. Ling, J.M., Zhang, L., Feng, Q.: Business process model alignment: an approach to support fast discovering complex matches. In: Mertins, K., Bénaben, F., Poler, R., Bourrières, J.-P. (eds.) Enterprise Interoperability VI: Interoperability for Agility, Resilience and Plasticity of Collaborations. Proceedings of the I-ESA Conferences, vol. 7, pp. 41–51. Springer, Heidelberg (2014)
7. Ling, J.M., Zhang, L.: Matching process model variants based on process structure tree. Chin. J. Softw. **26**(3), 460–474 (2015)
8. Klinkmüller, C., Weber, I., Mendling, J., Leopold, H., Ludwig, A.: Increasing recall of process model matching by improved activity label matching. In: Daniel, F., Wang, J., Weber, B. (eds.) BPM 2013. LNCS, vol. 8094, pp. 211–218. Springer, Heidelberg (2013)

9. Castelo Branco, M., Troya, J., Czarnecki, K., Küster, J., Völzer, H.: Matching business process workflows across abstraction levels. In: France, R.B., Kazmeier, J., Breu, R., Atkinson, C. (eds.) MODELS 2012. LNCS, vol. 7590, pp. 626–641. Springer, Heidelberg (2012)

10. Rosa, M.L., Dumas, M., Uba, R.: Business process model merging: an approach to business process consolidation. ACM Trans. Softw. Eng. Methodol. 22(2), 3–13 (2013)

11. Dijkman, R., Dumas, M., Van, D.B., Kaarik, R., Mendling, J.: Similarity of business process models: metrics and evaluation. Inf. Syst. 36(2), 498–516 (2011)

12. EMF Compare. http://www.eclipse.org/emf/compare. Accessed 2015

13. Rubin, J., Chechik, M.: Combining related products into product lines. In: Zisman, A., Lara, J. (eds.) FASE 2012. LNCS, vol. 7212, pp. 285–300. Springer, Heidelberg (2012)

14. Van den Brand, M., Protić, Z., Verhoeff, T.: Fine-grained metamodel-assisted model comparison. In: Proceedings of the International Workshop on Model Comparison in Practice, pp. 11–20 (2010)

15. Kolovos, D.S.: Establishing correspondences between models with the epsilon comparison language. In: Paige, R.F., Hartman, A., Rensink, A. (eds.) ECMDA-FA 2009. LNCS, vol. 5562, pp. 146–157. Springer, Heidelberg (2009)

16. Mehra, A., Grundy, J., Hosking, J.: A generic approach to supporting diagram differencing and merging for collaborative design. In: Proceedings of the IEEE/ACM International Conference on Automated Software Engineering, pp. 204–213 (2005)

A Management Tool for Distributed Heterogeneous Process Logs

Gui-yuan Yuan[1], Qing-tian Zeng[1,2(✉)], Hua Duan[3],
Fa-ming Lu[1,4], and Chang-hong Zhou[1]

[1] College of Information Science and Engineering,
Shandong University of Science and Technology, Qingdao 266590, China
1105793790@qq.com, {qtzeng,fm_lu,zhouchanghong}@163.com
[2] College of Electronic Communication and Physics,
Shandong University of Science and Technology, Qingdao 266590, China
[3] College of Mathematics and System Science, Shandong University of Science and Technology,
Qingdao 266590, China
[4] The Key Laboratory of Embedded System and Service Computing,
Ministry of Education, Tongji University, Shanghai 200092, China

Abstract. Process logs present the characteristics of distribution and heterogene in today's process aware information systems, which causes lots of difficulties in log management and integration. To address this problem, a management tool for distributed heterogeneous process logs is designed and developed. The tool supports the sharing of cross-organizational processes logs under privacy protection, and provides functional operations such as integration of distributed heterogeneous process log files, format standardization of heterogeneous process logs, visual presentation of case trajectory, clustering analysis of process cases attributes, pre-processing of process logs and so on. Compared with existing management tools for process logs, this tool is capable of facilitating the horizontal and vertical integration of distributed and heterogeneous logs, which has benefit the privacy protection of cross-organizational business processes logs and cross-organizational business process mining.

Keywords: Process mining · Event logs · Log integration

1 Introduction

Process mining [1, 2] aims to extract useful knowledge for process improvement from the process logs. However now the rich information system logs today present the characteristics of distributed and heterogeneous. Therefore it is generally required to use some process logs management tools to manage these distributed heterogeneous process logs.

XES (eXtensible Event Stream) [3] based on the XML format is widely used as the factorial standard in process log storage. It not only has a standard notation to store time, resources, activities and other properties, but also can add custom data elements to XES.

© Springer Science+Business Media Singapore 2016
J. Cao et al. (Eds.): PAS 2015, CCIS 602, pp. 79–86, 2016.
DOI: 10.1007/978-981-10-1019-4_7

In 2010, this format was accepted by IEEE Task Force on Process Mining. This paper will be based on XES as the standard of log format and heterogeneous logs.

As to process mining tools, Disco ProM [4] and Fluxicon [5] are widely used.

Fluxicon Disco which developed by Holland Fluxicon company can provide the automatic discovery of business process model, process log filtering, project management and heterogeneous process log format standardization functions, but for the ordinary users, the software is limited to the number of the process logs, and is not conducive in dealing with large process logs, and Disco dose not achieve the integrated function about distributed process logs. ProM is an open source process mining suit, which can be used for studying process mining cases. ProM supports the integration of log files in XES format, but it can't be integrated directly with the integration of heterogeneous logs, and it is not conducive in the integration of user log files with less integrated approach. XESame and ProMimport are two process logs management tools which are commonly used, XESame can import data from database table, and use data to create XES file. When users want to import data use XESame, users need to set database driver and database address, it is very difficult for user. ProMimport also is a process logs management tool, users can use it to create the logs file in MXML format, but it can't create XES logs file.

To this end, this paper develops a distributed and heterogeneous process logs management tool, which is based on the above advantages and disadvantages of the system also overcome some shortcoming of existing software. This paper introduces the software architecture and basic functions and tools, and gives the summary and prospect of this tool.

2   Software Architecture and Function Design

2.1   Software Architecture

The logical architecture of the distributed heterogeneous process logs management tool is shown in Fig. 1. According to the architecture, the system is divided into three layers, data layer, application layer and client layer.

Data layer: it is mainly composed of the database and the process log files, which are the process logs management tools need to be processed. The process data are accessed through JDBC and file access interface.

Application layer: this layer includes control layer and basic service layer. The control layer mainly deals with the process logs processing request, and calls the basic service layer class to process the request and the response. The basic service layer is composed of several classes, and provides service for client layer to process log data.

Client layer: this layer provides users with application program interface, and users can achieve the integrated function of cross-organizational process logs, distributed logs, heterogeneous process logs, case trail, process logs analysis, process data filtering, data visualization, data clustering, and so on.

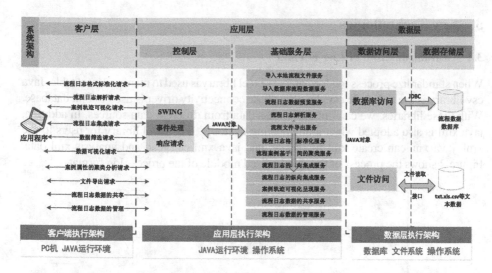

Fig. 1. System architecture diagram

2.2 Software Function

The proposed management tool mainly includes four functional modules, which are log preprocessing, distributed log integration, process log privacy, and management and process log data. Among them:

Log preprocessing: for the original data in the text file and data tables in the database, the tool can perform data filtering, data visualization, log data analysis, etc., and also can convert TXT, XLS, CSV or other text files and data tables in the database to XES format standard process log files.

Integration of distributed log: log integration is used to integrate multiple process logs into one log for further analyze. It can be divided into horizontal integration and vertical integration. Horizontal integration is based on the horizontal relation between the case attributes in different logs. For the two logs which have many attributes in common, vertical integration extends the properties of two logs to an attribute union (or attributes need to be extended and deleted that user specifies). After extension, the two logs have exactly the same properties, and are merged into a new log in the end.

Sharing and protection of the process log privacy: for cross-organizational business processes, the exchange of data between different organizations needs to be carried out. The process log data can share the process data of the organizations with other users, and the data can be shared through the process log data management.

The intelligent analysis of process log data: the visualization of case trail is the analysis of the order of the XES process log case, which makes users have an intuitive understanding about the order execution, and presents the trail of the case in a way of trail diagram; process cases of cluster analysis based on time is the case in process log, it can be clustered by some attributes data such as time, complete the cluster analysis.

3 Key Technologies

3.1 Standardization of Log Format

When standardize process log format, java excel library is used to manage Excel table. Java csv library is use to handle csv. Java data base connectivity driver is used to link database. With these libraries, we can import process data from different data sources. In addition, java xml is also adopted when implementing the tools. It is a tool library for users to use xml, java xml can create xml file easily, and java xml can read and analysis xml data. Figure 2 shows the process of the standardized module of the process log format.

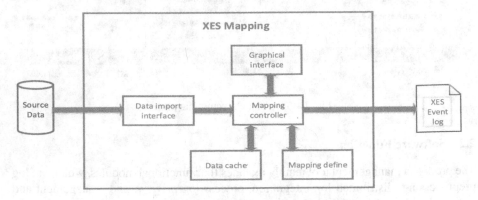

Fig. 2. The format standardization module of the tool

3.2 Integration of Distributed Logs

In distributed log integration module, two kinds of integration, i.e., horizontal and vertical integration of distributed log, are introduced. It uses process log data and the integration rules (manually set by user) to merge logs file in first, and then used the

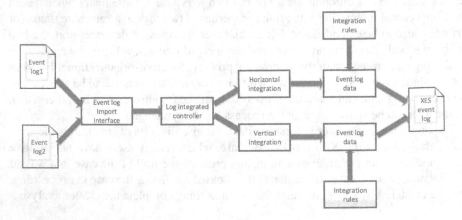

Fig. 3. Distributed log integrated module

standardized module of the process log format to create XES logs file. Figure 3 show the process of distributed log integration module.

4   Tool Implementation

4.1   Log Preprocessing Module

Log preprocessing includes filtering of raw data, visualization of raw data, parsing of the XES log, and the standardization of the process log format:

The standardized module of the process log format: standardized treatment for the process data of the user heterogeneous process log file and the process log database, making them the standard process log file of the XES format. Figure 4 shows the log source files and log files after standardized; XES log analysis of the storage module: parsing the XES log file which users import, presenting information which the XES file contains to users, exporting these data information to users.

Raw data visualization modules: presenting process logs or process database in the data sheet and original data in form of histograms or pie charts to the user, the user only needs to specify a column in the data flow; Filtering of the original data: the user can filter the data in the process log, and remove the unnecessary process data.

4.2   Integration Module Distributed Logs

Distributed log integration includes two modules, the horizontal integration of distributed log data, and the vertical integration of distributed log data:

Horizontal integration module of the distributed log data: one log records the process information of orders, including order number, customer name, order time, completion

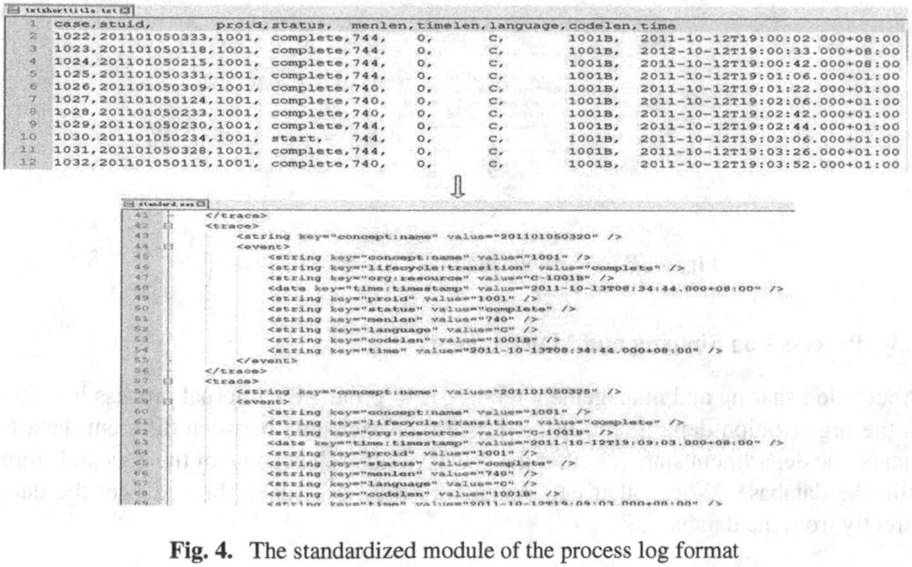

Fig. 4. The standardized module of the process log format

time and other information. Another log records the information of delivery, including record of ID, delivery order, delivery association number, delivery time and other information. The two logs can establish the main foreign key relationship through the common property "order number", to obtain an integrated log which has a description of order delivery records and the corresponding order completed information. This operation is called horizontal integration. Figure 5 shows the distributed log source files and log files after horizontal integration. Vertical integration of distributed log data: for example, there are two order tables. Both of them store the order number, the customer's name and other information. When the two process logs need to be integrated, you can specify which columns need to be integrated, and then through the vertical integration of distributed log data module.

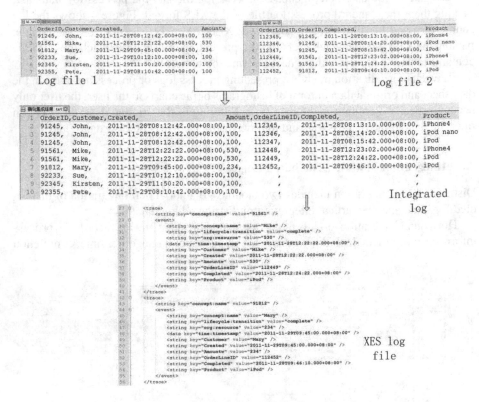

Fig. 5. Horizontal integration of distributed log data

4.3 Process Log Sharing and Management

Process log sharing and management module is to protect the internal process log data in the organization department. In the process of exchange between different departments, the department staff can filter the privacy data, select some of the data and store it in the database. When other departments need process data, they can get the data directly from the database.

4.4   Intelligent Analysis of Process Log Data

For the XES process log, the case log is not easy to observe, and is not conducive for the process of mining personnel to have a global understanding about the case of the occurrence of process data. In order to solve the above problems, the trajectory visualization module is designed. The module can present the case trail of process data in the form of diagram. Users can grasp the case trail in the process log and analyze it by these diagrams. Figure 6 shows the case trajectory visualization results for XES format of the log file.

Process case based on the time of the clustering analysis module uses clustering algorithm based on K-means, so it can achieve the process of clustering with the time attribute, and users can dig the information which is contained in the case.

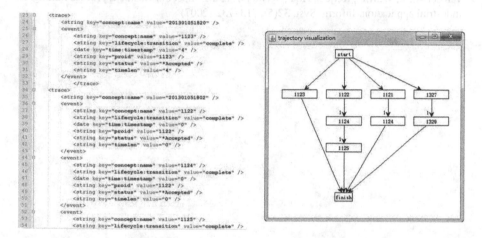

Fig. 6. Case trajectory visualization is presented to run the result diagrams

5   Summary

This paper introduced a distributed heterogeneous process log management tool, which supports the sharing of process logs under privacy protection, format standardization of heterogeneous process logs, log visualization, log filtering, horizontal and vertical integration of distributed logs. It can avoid the heavy manual work when management distributed heterogeneous process logs.

Acknowledgements.   This work was supported in part by NSFC (61472229, 61170079 and 61202152), by the Sci. & Tech. Development Fund of Shandong (2014GGX101035 and ZR2015FM013), by the Scientific Research Award Foundation for Outstanding Young Scientists of Shandong Province (BS2014DX013), by the open project of the Key Laboratory of Embedded System and Service Computing, Ministry of Education, Tongji University (ESSCKF201403), the Group-Star project of SDUST (qx2013113, qx2013354).

References

1. Bose, R.P.J.C., et al.: Dealing with concept drifts in process mining. Neural Netw. Learn. Syst. **25**(1), 154–171 (2013). Department of Mathematics and Computer Science Eindhoven University of Technology
2. Bannert, M., Reimann, P., Sonnenberg, C.: Process mining techniques for analysing patterns and strategies in students' self-regulated learning. Metacogn. Learn. **9**(2), 161–185 (2014)
3. van Dongen, B.F., Shabani, S.: Relational XES: data management for process mining. In: CAiSE Forum 2015, pp. 169–176 (2015)
4. Schmitt, B.: Process mining with process observer and Fluxicon Disco[EB/OL]. http://scn.sap.com/community/bpm/blog/2014/11/17/process-mining-with-process-observer-and-fluxicon-disco
5. van der Aalst, W.M.P., Reijers, H.A., Weijters, A.J.M.M., et al.: Business process mining: an industrial application. Inform. Syst. **32**(5), 713–732 (2007)

Cloud Workflow Applications

An Effective Energy Testing Framework for Cloud Workflow Activities

Zhou Zhao[1], Xiao Liu[1(✉)], Juan Li[2], Kepi Zhang[1], and Jin Liu[2]

[1] Shanghai Key Lab for Trustworthy Computing, East China Normal University,
Shanghai, China
xliu@sei.ecnu.edu.cn
[2] State Key Lab for Software Engineering, Wuhan University, Wuhan, China

Abstract. Cloud computing as the latest computing paradigm has shown its promising future in business workflow systems facing massive concurrent user requests and complicated computing tasks. With the fast growth of cloud data centers, energy management especially energy monitoring and saving in cloud workflow systems has been attracting increasing attention. It is obvious that the energy for running a cloud workflow instance is mainly dependent on the energy for executing its workflow activities. However, existing energy management strategies mainly monitor the virtual machines instead of the workflow activities running on them, and hence it is difficult to directly monitor and optimize the energy consumption of cloud workflows. To address such an issue, in this paper, we propose an effective energy testing framework for cloud workflow activities. This framework can help to accurately test and analyze the baseline energy of physical and virtual machines in the cloud environment, and then obtain the energy consumption data of cloud workflow activities. Based on these data, we can further produce the energy consumption model and apply energy prediction strategies. Our experiments are conducted in an OpenStack based cloud computing environment. The effectiveness of our framework has been successfully verified through a detailed case study and a set of energy modelling and prediction experiments based on representative time-series models.

Keywords: Cloud computing · Energy testing · Business workflow · Time-series model

1 Introduction

Cloud computing is a new computing paradigm based on the development of distributed computing [1], grid computing [2], peer to peer computing [3] and many other traditional computing technologies. Cloud data center is the heart of cloud computing and there are many key technologies for the management of cloud data center such as cloud data storage [4], cloud data management [5], virtualization [6, 7], energy saving technologies and so on and so forth. Through the integration of these technologies, the ultimate goal of cloud computing is to provide enterprises and personal users with cheap and easy-access cloud services. In recent years, with fast development of cloud computing and

© Springer Science+Business Media Singapore 2016
J. Cao et al. (Eds.): PAS 2015, CCIS 602, pp. 89–105, 2016.
DOI: 10.1007/978-981-10-1019-4_8

the growth in the size of cloud data centers, the investment on cloud infrastructures increased drastically. As a result, energy consumption becomes the major running cost for cloud computing and it also brings the serious problem to the environment [27].

According to the public statistics, the power consumption of data centers in USA is 91 billion KWH in 2013, which is equal to the power generated by a total of 34 power plants each with an annual production of 500 trillion watts, and also equal to 2 times the annual power consumed by the city of New York. It is estimated that by the year 2020, the yearly power consumption of data centers will reach to 140 billion KWH, which is equal to the power generated by 50 powers plants as mentioned above, and costs 13 billion US dollars while produces 150 billion tons of Carbon Dioxide [8].

Table 1 shows the power consumption of data centers in China.

Table 1. Annual power consumption of data centers in China

Year / One hundred million KWH	2009	2010	2011	2012	2013	2014
Power consumption	1070	1230	1412	1634	1897	2210
Increment	13.9%	14.9%	14.8%	15.7%	16.1%	16.5%

Energy consumption is the major running cost for cloud infrastructures and hence energy becomes the key in the management of cloud data centers [28]. In recent years, intensive studies on the issue of energy saving for cloud infrastructures were carrying out by both academia and industry, the idea of green computing and effective energy-saving strategies have been proposed [29]. As a typical business software service in the cloud, the energy saving issue in cloud workflow systems has also attracted the attention of many researchers. Business Process Management (BPM) [9] integrates workflow technology to enterprise's business management and realizes information transfer, data synchronization, business monitoring and process optimization through the Internet. In the early stage, most BPM systems are centralized systems which have inherent problems on system performance, reliability and scalability. As the latest computing paradigm, cloud computing enables the enterprises to get access to unlimited computing, storage and information services in an on-demand fashion. Therefore, cloud workflow systems can have the ability to dynamically scale according to the workload, and hence they are becoming increasingly popular in the BPM area especially for those systems facing massive concurrent user requests and complicated computing tasks. From the perspective of business process, the energy consumed for running a business workflow is mainly dependent on the total energy consumed by the execution of workflow activities. However, current studies on the energy management of cloud workflows, such as on the modelling and prediction of energy consumption, and energy-aware workflow scheduling, the monitoring objectives are cloud virtual machines instead of workflow activities running on them. Since the major advantage of workflow systems is the separation between business logic and business activities so as to provide general support to different business processes, the monitoring objectives should be workflow activities themselves. In fact, we can effectively monitor and optimize the energy consumption of cloud workflows only after accurate testing, modelling and prediction of the energy consumption of cloud workflow activities. However, it is easy to monitor the energy consumption of a physical machine, but it is much more complicated for energy monitoring in a distributed virtualized cloud computing environment.

To address such an issue, in this paper, we propose an effective energy testing framework for cloud workflow activities. This framework can obtain the baseline power of physical machine and virtual machine in a cloud environment, and then achieve the accurate figures of power required for running workflow activities. Based on such a result, we can further make effective modelling and prediction. With an OpenStack based private cloud environment, a detailed case study and a set of time-series models based modelling and prediction results have been demonstrated to prove the effectiveness of our energy testing framework. The work presented in this paper can provide the fundamentals for the future investigation on the energy consumption of entire cloud workflows.

The remainder of this paper is organized as follows. Section 2 introduces the work related to the energy management for cloud workflow systems. Section 3 proposes our energy testing framework for cloud workflow activities. Section 4 demonstrates the effectiveness of our framework through a detailed case study and a set of modelling and prediction results. Finally, Sect. 5 concludes this paper and points out some future work.

2   Related Work

Cloud computing becomes a popular solution for decreasing the running cost of customer's IT systems. However, with the fast expansion of cloud data centers, expensive energy consumption becomes increasingly serious which will hinder the growth of cloud computing. Gu et al. conducted a survey on the energy consumption of China's data centers and the results show that the energy efficiency is relatively low [10]. Chen et al. investigated energy-saving techniques for virtualized cloud data centers, especially addressing the energy waste in cooling systems due to the uneven distribution of heat in data centers [11]. The work in [12] introduced the method for energy testing and summarized the current progress on the energy management from both the virtualization level and platform level.

There are some simple energy models for cloud computing in the literatures. Jayant et al. investigated the energy consumption of software service, storage service and platform service in the cloud through data transfer between public clouds and private clouds [13]. They presented and compared the energy consumption of computers, routers and data centers in detail but no energy saving solutions were proposed. The work in [14] employed CPU performance counter and system utilization rate and applied multiple variates regression and nonlinear regression models. The authors analyzed the effectiveness of different models with various parameters and proposed the energy models which suit the cloud data center infrastructures. Currently, many existing literatures adopt multiple variates regression models based on CPU utilization rate, RAM utilization rate, I/O speed and other variates to predict the energy consumption of physical servers. As many studies have shown that CPU can consume about 60 % of the total power when it reaches to the maximum utilization rate, and hence as a major energy component, a set of models were proposed based on the CPU power during the peak and sleep status, and then used it for the energy prediction of physical servers.

Currently there are many commercial public clouds such as Amazon Web Services (http://aws.amazon.com/), Windows Azure (http://azure.microsoft.com/), and Ali Yun (http://www.aliyun.com/). Recent years also witness the fast growth of open source cloud computing platforms such as OpenStack (http://www.openstack.org/), CloudStack (https://cloudstack.apache.org/) and Cloud Foundry (https://www.cloudfoundry.org/). Among them OpenStack is the most popular one and many people believed it is like "Linux" in the area of operating systems. In OpenStack, the component named Ceilometer (http://wiki.openstack.org/Ceilometer) is designed to provide the metering and monitoring services. With Ceilometer, system managers can obtain and store the real-time and historic data for many metrics for CPU, RAM, Disk and network, so that metering and monitoring functions can be fulfilled. Given those collected data, data analysis based on various techniques can be applied for energy prediction and saving. However, in the current version of Ceilometer, no energy monitoring mechanism has been developed yet.

Cloud computing based business process management systems and cloud workflow systems are becoming a hot topic in both academia and industry [15–17]. Many major business system developer companies have provided their business process management software such as IBM Business Process Manager on Cloud [17, 18] and Microsoft Windows Workflow Foundation [20]. SAP Business Workflow [19] provides support for cloud services and is planning to release a full cloud based version. In academia, there are already many cloud workflow systems dedicated to scientific computing and distributed computing such as Kepler [23], Pegasus [24] and CloudBus [22], and also include SwinFlow-Cloud [21, 25] which is a cloud workflow system dedicated to massive concurrent business workflow instances. Recently, in the area of cloud workflow scheduling, many researchers employed heuristic algorithms where energy is regarded as one of the optimization objectives together with response time, throughput, cost and other QoS (Quality of Service) dimensions [30, 31, 32]. However, those energy-aware scheduling algorithms are still working on the energy saving at the virtual machine level not directly for workflow activities. The major problem here is the lack of effective energy modeling and prediction for cloud workflow activities.

3   An Energy Testing Framework for Cloud Workflow Activities

In this paper, we propose an energy testing framework for cloud workflow activities and implement it in an OpenStack based private cloud environment so that we can obtain the energy consumption data directly through hardware power tester. We can also implement the framework in public cloud environment but in such a case we need to change the hardware power tester to its software counterparts such as Joulemeter (http://research.microsoft.com/en-us/projects/joulemeter/) and Powertop (https://01.org/zh/powertop), and all the others are the same. However, it should be noted that the energy consumption data collected directly by hardware power testers is much more accurate than software meters which are mainly based on models and algorithms. Therefore, we should employ hardware meters whenever possible. Figure 1 shows the overview of our framework. It can be seen that the testing framework consists of four layers, from bottom to top, including Physical Layer, Resource Management Layer, Virtual Layer and Application Layer.

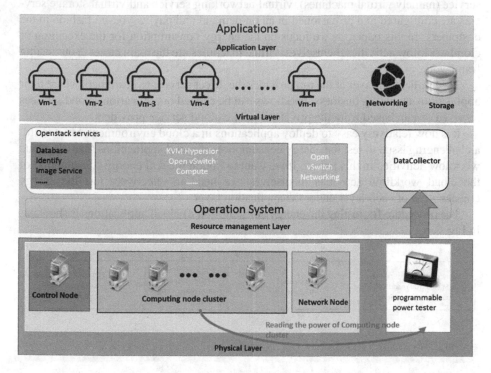

Fig. 1. Energy testing framework

The Physical layer consists of two parts, one is the physical resource part including such as the control node, the computing node cluster and network node to provide the infrastructure for the cloud environment, and the other is the programmable power tester which can be used to directly meter the target and store its real-time power data through programming.

Above the physical layer is the resource management layer which consists of three parts including the operation system, the cloud resource management service and the DataCollector. The operation system manages the physical resources, the cloud resource management service (namely OpenStack services in our framework) manages the virtualized resources, and the DataCollector is an implementation of the programming interface of the power tester which can meter and store the real-time power data. As shown in the figure, the components in the resource management layer are corresponding to the components in the physical layer. Specifically, the control nodes manage the metadata of the cloud environment and provide the identity service and also the image service for virtual machines; the computing node cluster provides the computing service and manages virtual machines; the network nodes provide network service and manage the complicated virtualized network. Besides OpenStack, there are also other cloud resource management platforms including open-source platforms such as CloudStack and Cloud Foundry, and also commercial platforms such as AWS and Windows Azure.

The virtualization layer provides virtualized resources. Virtualized computing service (namely virtual machines), virtual networking service and virtual storage services are configured and provisioned in an on-demand and pay-as-you-go fashion to the customers. In this paper, as we focus on the energy consumption for the execution of cloud workflow activities themselves, virtual machines are the main target in our testing framework.

The application layer at the top provides the direct service to customers where their applications including business workflows can be carried out by various cloud services reserved or purchased from the private or public cloud service providers.

It is now relatively easy to deploy applications in a cloud environment, but it is still a challenging issue to test the energy consumption of cloud applications (namely cloud workflow activities in this paper). If we want to obtain the real energy consumption of the cloud workflow activities themselves, we need to effectively test the energy consumption of each layer under the application layer.

The procedure for testing the energy consumption of cloud applications is shown as in Fig. 2.

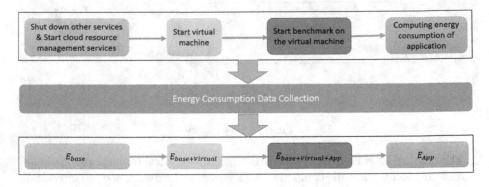

Fig. 2. The procedure for energy testing of cloud workflow activities

The whole testing procedure centers on the collection of energy consumption data at different layers. In order to test the baseline power at different layers, we need to gradually start the services layer by layer. Specifically, according to the testing framework above, to test the energy of a cloud workflow activity running on a cloud virtual machine, our testing procedure is as follows: (1) all services need to be shut down except the cloud management services (e.g. OpenStack service in this paper); (2) meter the energy consumption denoted as E_{base} which is the baseline energy consumption for physical severs with cloud management services; (3) apply and start the cloud virtual machine; (4) when the virtual machine is stable, meter the energy denoted as $E_{base+Virtual}$

which is the baseline energy consumption after virtual machine is running; (5) start the testing objective namely the cloud applications or cloud workflow activities in this paper; (6) meter the energy consumption $E_{base+Virtual+App}$ which is the total energy consumption after the cloud workflow activities are up and running on the cloud virtual machines. (7) the last step is to calculate the energy consumption of the cloud workflow activities themselves using all the obtained energy consumption data.

Considering the extra power required during the start of the virtual machines and cloud applications (detailed example will be shown in Sect. 4), all energy consumption data should be collected after the power is stable. According to different layers and the testing data, we can obtain the following equations:

$$E_{Virtual} = E_{base+Virtual} - E_{base};$$
$$E_{App} = E_{base+Virtual+App} - E_{base} - E_{Virtual}$$
$$= E_{base+Virtual+App} - E_{base} - (E_{base+Virtual} - E_{base});$$

and finally we can get that: $E_{App} = E_{base+Virtual+App} - E_{base+Virtual}.$

Given our energy testing framework, we can obtain the energy consumption data for cloud workflow activities. After that, different modelling and prediction strategies can be applied to model and predict the energy consumption of cloud workflow activities. These modelling and prediction results will be of great importance for cloud workflow resource management, cloud workflow scheduling, and many other decision tasks for energy saving in a cloud workflow systems. In the next section, we will present a set of comprehensive experimental results to evaluate the effectiveness of our energy testing framework.

4 Experiment

Section 3 proposes our energy testing framework. We divide the total energy consumption into different layers and then measure the energy consumption layer by layer, and finally we can get the energy consumption of the cloud workflow activities. This chapter provides comprehensive experimental results to evaluate the effectiveness of our energy testing framework.

4.1 Experimental Environment

Our experimental environment is an OpenStack based private cloud. Table 2 shows the configuration of the physical layer in the cloud environment.

Table 2. Configuration of the physical cluster

Info / Nodes	CPU	RAM	Number of Network Interface Adapter for each computer	Number of computer
Control Node	4 cores	8G	1	1
Network Node	4 cores	8G	4	1
Computing cluster	8 cores	32G	3	3

In the physical layer, we use the HP-8713 (more information about HP-8713 please visits http://www.china.cn/qtdiangongyiqiyib/3394492756.html) as the programmable power tester. HP-8713 is a real-time power meter which is able to collect the power data of the object and display the value of the voltage, current and power through the monitor screen. In addition, HP-8713 provides programmable interface. According to the interface protocol, we can implement the interface to collect and store the power data in a text format with adjustable frequency. As shown in Fig. 1, DataCollector in the Resource Management Layer is a data collector program implementing the interface provided by HP-8713.

In our cloud environment, we use Centos 7 as the operating system for physical servers and OpenStack (Juno release version) as the cloud computing resource management service. The control node is mainly installed with the authorization service, database service and the Dashboard; the computing cluster nodes mainly provide the virtualized computing service where computing virtualization service and some services for communication with other services are installed; the network node is installed with the Neutron module which provides the networking service. We can configure the settings of the virtual machines according to resource demand and produce virtual machines based on customer provided machine images to achieve different software environments. In our experiments, we assign 2 cores and 4G RAM for each virtual machine. The operation system of each virtual machine is Centos 6.6.

4.2 Case Study

Here, we present two case studies. The first one demonstrates the detailed results for testing the energy consumption of a cloud workflow activity on a single virtual machine. The second demonstrates the results with the same activity but with multiple instances and multiple virtual machines to simulate concurrent user requests. In addition, two different resource management strategies are also implemented for comparison purpose. In our case study, we use prime95 as the benchmark application. Prime95 is a free application for Mersense Prime search provided by the Great Internet Mersense Prime (GIMPS, http://www.mersenne.org/), which is a distributed computing project based on Internet. It is a typical CPU-intensive application representing cloud workflow activities with complicated computations. Other type of applications such as data intensive and I/O intensive can be implemented in a similar fashion.

4.2.1 Application Energy Testing on a Single Virtual Machine

In order to clearly illustrate the energy testing process of a single application, we first deploy the application on a single virtual machine and test its energy consumption value. The relationship between energy consumption and the power can be expressed as a simple formula: $E = p*t$, that is, the energy consumption can be calculated by power multiplied by the duration time of the application. For most workflow activities, the duration time of the activity is predictable [27]. Therefore, as shown in the following figures, we focus on the power data directly in our experiment.

After the cloud environment is ready, the first Phase is to start the cloud services and record the power data after the power becomes relatively stable. We can run the system for a while to obtain a sufficient size of dataset and then go to the next step. Phase 2 is to start the virtual machine. As can be seen in Fig. 3, the power has a big surge during the start process and then become stable after the virtual machine starts completely. Phase 3 is to start prime95. The power data acquisition program namely DataCollector is always running during the whole process. The test program will stop when it reaches some stopping conditions and then we will find the power drops significantly. We repeat the tests for many times and at each time only start one single virtual machine in the cloud environment. Figure 3 shows the data collected during the process and Fig. 4 depicts the differences of the power obtained at each Phase.

Specifically, Phase 1 in Fig. 3 represents the power before starting the virtual machine but after starting the cloud service; Phase 2 represents the power of starting the virtual machine; Phase 3 is the power after starting the testing application; Phase 4 shows the power after closing the testing application.

Big surges take place at the turning point between two adjacent phases as shown in Fig. 3. Furthermore, Fig. 4 depicts the difference of the power between different phases.

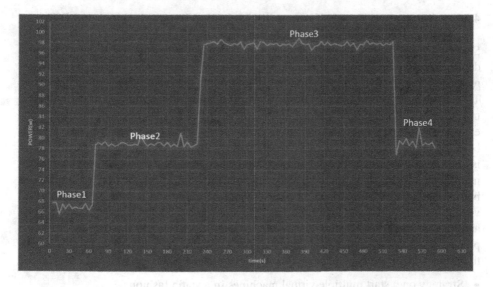

Fig. 3. The power time series of a single virtual machine

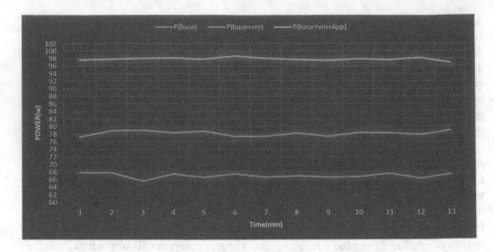

Fig. 4. Power difference between phases

We can calculate the average of the time-series data and get the power value: $P_{base+vm} = 78.13w$, $P_{base+vm+App} = 97.87w$. According to discussion in Sect. 3, we can obtain that the energy consumption of application can be calculated by $E_{App} = E_{base+vm+App} - E_{base+vm}$. Similarly, given the relationship between power and energy, we can get that the average power of the testing application is: $P_{App} = P_{base+vm+App} - P_{base+vm} = 19.74w$. Its energy consumption can be easily calculated by multiple its average power with its running time.

4.2.2 Application Energy Testing with Multiple Virtual Machines and Different Resource Management Strategies

Section 4.2.1 describes the method for testing the energy consumption of the application deployed on one single virtual machine. For distributed application services or concurrent user requests, many applications need to be deployed on multiple virtual machines. Thus there is a need to monitor the power of the applications running in many virtual machines. In addition, in the cloud environment, different resource management strategies are employed. For example, provisioning all required resources in a static fashion before workflow execution is one typical strategy. In contrast, another typical strategy is to provision resources in a dynamic fashion where new resources are provided incrementally according to the increase of the resource demand. In this section, we implement both strategies for comparison purpose. Specifically, the static strategy is to start all virtual machines first and then start the application; the dynamic strategy is to start one application immediately after starting one virtual machine, and then repeat the same process for multiple virtual machines. We do the tests for these two strategies to observe the impact of different strategies on the energy consumption.

- Strategy one: start multiple virtual machines in a static fashion

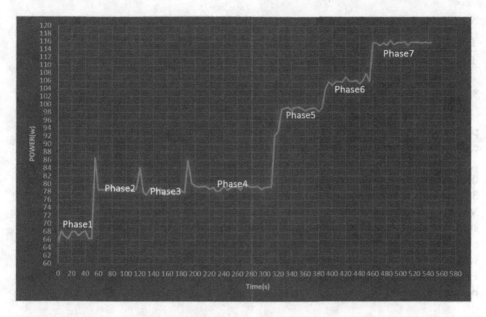

Fig. 5. The power time series where virtual machines are provisioned in a static fashion

As shown in Fig. 5, Phase 1 represents the power starting the cloud service which is the baseline power; Phase 2 represents the power when VM1 is active and stable; Phase 3 represents the power when VM2 is active and stable; Phase 4 represents the power when VM3 is active and stable; Phase 5 represents the power starting prime95 deployed on VM1; Phase6 represents the power starting prime95 deployed on VM2; Phase 7 represents the power starting prime95 deployed on VM3. As can be seen in the figure, the power has a big increase when starting the first VM. But after that, the power remains almost stable when starting the other two VMs except for a short-term surge at the beginning of each Phase. This is a very strange finding as almost none extra power is produced when new virtual machines are provisioned. Without investigating the detailed internal management mechanism for virtual machines in OpenStack, we cannot explain the reason for such a finding. However, it can be regarded at least as a real example to show the big difference between the energy management of traditional computing environments and the new cloud computing environments.

Table 3 shows the average power during each Phase. After starting the physical machine and cloud service, the baseline power of the cluster is 67.13 W, that is, $P_{base} = 67.13w$. Phase 2, 3, and 4 represents respectively the cluster power of starting three VMs. The results illustrate that after starting the first VM, the power has a bit change when starting the remaining two VMs. And the power value reaches to $P_{base+vm} = 79.17w$ after starting all the VMs. And then start the application deployed on these VMs one by one. The cluster power reaches to $P_{base+vm+App} = 115.66w$ after successfully starting all applications. Finally, we can get the power of the applications deployed on the three VMs by formula $P_{App} = P_{base+vm+App} - P_{base+vm} = 115.66w - 79.17w = 36.49w$.

Table 3. Average power where virtual machines are provisioned in a static fashion

Phase	Phase 1	Phase 2	Phase 3	Phase 4	Phase 5	Phase 6	Phase 7
Power(w)	67.13	78.06	78.73	79.17	98.14	105.98	115.66

- Strategy two: start multiple virtual machines in a dynamic fashion

This test is to start the application immediately after starting one VM and repeat the process for multiple VMs in the OpenStack cloud environment. Figure 5 shows the change of the power during the whole test.

As shown in Fig. 6, Phase 1 represents the power starting the cloud service which is the baseline power; Phase 2 represents the power when VM1 is active and stable; Phase 3 represents the power starting prime95 on VM1; Phase 4 represents the power when VM2 is active and stable where there is almost no change; Phase 5 represents the power starting prime95 on VM2. Comparing the power of phase 5 and phase 4, there is an obvious increase between them due to the application. Phase 6 represents the power when VM3 is active and stable. Similar as starting VM2, the power has little change in Phase 6. Phase 7 represents the power starting prime95 on VM3 and there is an obvious power surge.

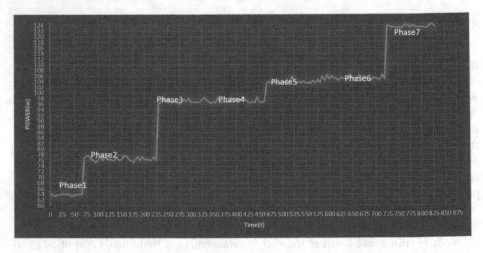

Fig. 6. The power time series where virtual machines are provisioned in a dynamic fashion

Table 4 shows that the average power of Phase 2 is 76.71 when the system starts only one VM. The increase of the power from Phase 3 to Phase 4 and from Phase 5 to Phase 6 can be seen as the power consumption of VM, thus the power value of three VMs add the base infrastructure can be calculated by formula $P_{base+vm} = 76.71 + 97.14 - 96.93 + 105.24 - 103.95 = 78.21w$. The average power of phase 7 can be serve as the whole power consumption $P_{base+vm+App} = 123.44w$, finally, the power consumption of application is equal to $P_{App} = P_{base+vm+App} - P_{base+vm} = 123.44w - 78.21w = 45.23w$.

Table 4. Average power where virtual machines are provisioned in a dynamic fashion

Phase	Phase 1	Phase 2	Phase 3	Phase 4	Phase 5	Phase 6	Phase 7
Power(w)	63.85	76.71	96.93	97.14	103.95	105.24	123.44

The comparison results between two different resource management strategies show that the energy consumption of virtual machines in an OpenStack cloud environment is very different from the physical machines. Increasing the number of standby VMs does not cause a great change of energy consumption. Only at the time of starting the first virtual machine, the total power of the cluster has a significant increase. However, deploying and starting the application on the VMs always cause a big surge of the energy consumption, which is no different with traditional centralized application service. The comparison results show that the influence of different resource provisioning strategies on the baseline power is trivial, but it has an obvious effect on total power of applications. Under the same experimental environment, the static strategy produces a total of 36.49 W for all applications while the dynamic strategy produces a total of 45.23 W for all applications. According to such a result, we may infer that in an OpenStack cloud environment, the static resource provisioning strategy which starts all virtual machines upfront can achieve better energy saving effect than the dynamic resource provisioning strategy which starts virtual machines one by one according to the change of demand. This conclusion is contradictive with the common thought that increasing computing resources in an on-demand dynamic fashion can save the energy. However, it is important to note that in this paper we use the OpenStack cloud environment as an example, as to whether this conclusion can be applied to other cloud environments still needs further testing and validation.

Through the results of these experiments, we can observe that resource management and energy consumption in a cloud environment can be very different from the common experiences in other traditional computing environments. Therefore, having an effective energy testing framework for cloud workflow activities is very necessary for the energy monitoring and saving in cloud workflow systems.

4.3 Time-Series Based Energy Modelling and Prediction

Based on the energy consumption data collected in the above experiments, in this section, we will employ representative time-series models to demonstrate the modelling and prediction of the power of cloud workflow activities. From the modelling and prediction point of view, there are two kinds of activities: one is known and the other is unknown. For the known activity which has historical power data in the system, the models, such as ARMA (Auto-Regression and Moving Average) and ES (Exponential Smoothing), can be used to model and predict the power. For the unknown activity which is a new-coming in the system, we can segment the existing historical power data into several power patterns by time-series segmentation models, such as sliding windows model and bottom-up model. Afterwards, we can find the most similar pattern for the activity based on power pattern matching, and then the power for the unknown activity can be predicted by using the matched power pattern. Same as in our energy testing framework, we are able to record the power series of the cloud system when running the activity. The power series of the activity can be obtained by the power series of the system minus the baseline power series. Finally, the energy consumption can be calculated by the product of the power series of the activity and its duration time. Please refer to [33] for more details about the time-series based energy modelling and prediction strategies.

In our experiments, taking ES model and sliding windows model as example, we will describe the detailed modeling and prediction process.

- ES model for known activities

For these known activities, the ES model can be trained by the historical power series data. According to the smoothing coefficient, the ES models can be divided into Simple Exponential Smoothing (SES), Double Exponential Smoothing model (DES) and others. In this paper, we only demonstrate the results with DES. Taking the example of Fig. 3 in which only one virtual machine is active, we obtain the initial power series when running the prime95, as seen in the Phase 3 of Fig. 3. The fitting result of DES model is described in Fig. 7(a) and (b) depicts that the relevant error between the predicted power and the real power is from−1.2 % to 1.2 %, which shows that the accuracy of DES model is very promising.

- Sliding windows model for unknown activities

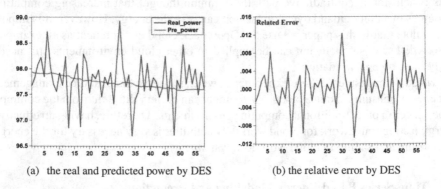

(a) the real and predicted power by DES   (b) the relative error by DES

Fig. 7. The fitting results of DES model

Section 4.2.2 has shown the power variation of cloud system under different resource provisioning strategies. In this section, we will take the first static strategy as an example. Given our energy testing framework, we can record the power series of every activity running in the cloud workflow system and create a database for historical power series. To apply the time-series segmentation method, we need to run the unknown application for a little while to get its small and incomplete initial power series. Then segment the initial power series into several power patterns using sliding windows model. Finally, valid the power pattern and obtain the real power pattern. The results of segmentation and validation are shown in Fig. 8. As depicted in Fig. 8, the initial power series has been segmented into 9 segmentations. It is obvious that there is a turning point in pattern 1 and 6 which can be seen as false patterns. After the validation process, there are 7 power patterns in the power series. The result is consistent with the 7 phases observed in Fig. 5, which illustrates the feasibility and veracity of sliding windows segmentation model. Given the database of historical power series, we can compare all existing patterns with the small and incomplete power series of the unknown activity and find out the most similar one to predict its future power series.

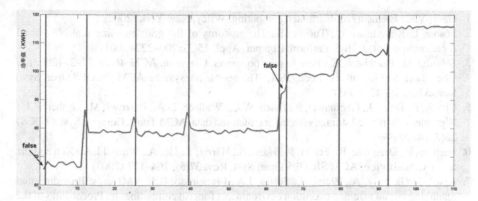

Fig. 8. Pattern segmentation and validation results of sliding windows model

5 Conclusion and Future Work

Cloud workflow systems become increasingly popular in the business process management for massive concurrent user requests and complicated computation tasks. However, with the fast growth of energy consumption in cloud data centers, the problem of energy monitoring and saving in cloud workflow systems attracted more and more attention. However, existing studies on cloud workflow energy consumption focus on virtual machines instead of cloud workflow activities running on them. Therefore, it is very difficult to model and predict the energy consumption of cloud workflows in such an indirect fashion. To address such a problem, this paper proposed an effective energy testing framework for cloud workflow activities where we can focus on the real energy consumption for cloud workflows themselves. Our framework can effectively test and analyze the baseline energy consumption of physical machines and virtual machines in a cloud environment, and then obtain the accurate energy consumption data for cloud workflow activities. Based on the data collected, we can make accurate energy modelling and prediction for cloud workflow activities. To demonstrate the effectiveness of our energy testing framework, with an OpenStack based private cloud environment, a detailed case study with the energy consumption under different resource management strategies and a set of modelling and prediction results based on representative time-series models were presented.

The work presented in this paper established a fundamental basis for the energy modelling and prediction of cloud workflow activities. In the future, we will investigate the energy consumption for an entire cloud workflow instance, and propose the energy modelling, prediction and saving strategies for cloud workflow systems.

Acknowledgement. The research work reported in this paper is partly supported by National Natural Science Foundation of China (NSFC) under No. 61300042, and Shanghai Knowledge Service Platform Project No. ZF1213.

References

1. Garg, V.K.: Elements of distributed computing. Wiley, New York (2002)
2. Foster, I., Kesselman, C., Tuecke, S.: The anatomy of the grid: enabling scalable virtual organizations. Int. J. High Perform. Comput. Appl. **15**(3), 200–222 (2001)
3. Schoder, D., Fischbach, K.: Peer-to-peer prospects. Commun. ACM **46**(2), 27–29 (2003)
4. Ghemawat, S., Gobioff, H., Leung, S.T.: The google file system. ACM SIGOPS Oper. Syst. Rev. **37**(5), 29–43 (2003)
5. Chang, F., Dean, J., Ghemawat, S., Hsieh, W.C., Wallach, D.A., Burrows, M., Gruber, R.E.: Bigtable: a distributed storage system for structured data. ACM Trans. Comput. Syst. (TOCS) **26**(2), 4 (2008)
6. Barham, P., Dragovic, B., Fraser, K., Hand, S., Harris, T., Ho, A., Warfield, A.: Xen and the art of virtualization. ACM SIGOPS Oper. Syst. Rev. **37**(5), 164–177 (2003)
7. Krsul, I., Ganguly, A., Zhang, J., Fortes, J.A., Figueiredo, R.J.: VMPlants: Providing and managing virtual machine execution environments for grid computing. In: Proceedings of the ACM/IEEE SC 2004 Conference on Supercomputing 2004, p. 7. IEEE, November 2004
8. Data Center Efficiency Assessment. http://www.nrdc.org/energy/files/data-center-efficiency-assessment-IP.pdf. Accessed 1st September 2015
9. van der Aalst, W.M., Ter Hofstede, A.H., Weske, M.: Business process management: a survey. In: van der Aalst, W.M., Weske, M. (eds.) Business Process Management, pp. 1–12. Springer, Heidelberg (2003)
10. Gu, L.J., Zhou, F.Q., Meng, H.: Research on Chinese energy consumption and energy efficiency of data center. Energy China **11**, 42–45 (2010)
11. Chen, H.: Energy Management Technology research in Virtualized Data Center. Beijing University of Posts and Telecommunications, Beijing (2012)
12. Ye, K.J., Wu, C.H., Jiang, X.H.: Power management of virtualized cloud computing platform. Chin. J. Comput. **06**, 1262–1285 (2012)
13. Baliga, J., Ayre, R.W., Hinton, K., Tucker, R.S.: Green cloud computing: balancing energy in processing, storage, and transport. Proc. IEEE **99**(1), 149–167 (2011)
14. Luo, L., Wu, W.J., Zhang, F.: Energy modeling based on cloud data center. J. Softw. **07**, 1371–1387 (2014)
15. Liu, X., Yuan, D., Zhang, G., Li, W., Cao, D., He, Q., Yang, Y.: Cloud workflow system functionality. The Design of Cloud Workflow Systems. Springer Briefs in Computer Science, pp. 19–25. Springer, New York (2012)
16. Liu, S.W., Kong, L.M., Ren, K.J.: A two-step data placement and task scheduling strategy for optimizing scientific workflow performance on cloud computing platform. Chin. J. Comput. **11**, 2121–2130 (2011)
17. Zhang, P., Wang, G.L., Xu, X.H.: A date placement approach for workflow in cloud. J. Comput. Res. Dev. **03**, 636–647 (2013)
18. IBM Business Process Manager on Cloud. http://www-03.ibm.com/software/products/zh/business-process-manager-cloud/. Accessed 1st September 2015
19. An Introduction to SAP Business Workflow. http://scn.sap.com/docs/DOC-31056. Accessed 1st September 2015

20. Milner, M.: A Developer's Introduction to Windows Workflow Foundation (WF) in .NET 4. http://msdn.microsoft.com/en-us/library/ee342461.aspx. Accessed 03 March 2014
21. SwinFlow-Cloud. http://www.ict.swin.edu.au/personal/dcao/. Accessed 03 March 2014
22. CloudBus Project. http://www.cloudbus.org/. Accessed 03 March 2014
23. Kepler Project. http://kepler-project.org/. Accessed 03 March 2014
24. Pegasus Project. http://pegasus.isi.edu/. Accessed 03 March 2014
25. Liu, X., Yang, Y., Yuan, D., Zhang, G., Li, W., Cao, D., He, Q., Chen, J.: The Design of Cloud Workflow Systems. Springer, Heidelberg (2012). ISBN 978-1-4614-1933-4
26. Liu, X., Ni, Z., Yuan, D., Jiang, Y., Wu, Z., Chen, J., Yang, Y.: A novel statistical time-series pattern based interval forecasting strategy for activity durations in workflow systems. J. Syst. Softw. **84**(3), 354–376 (2011)
27. Lv, T.W.: Deep analysis and outlook based on China Green Data Center. The World of Power Supply **12**, 6–8 (2012)
28. Guo, Y., Gong, Y., Fang, Y., Khargonekar, P.P., Geng, X.: Energy and network aware workload management for sustainable data centers with thermal storage. IEEE Trans. Parallel Distrib. Syst. **25**(8), 2030–2042 (2014)
29. Chase, J.S., Anderson, D.C., Thakar, P.N., Vahdat, A.M., Doyle, R.P.: Managing energy and server resources in hosting centers. ACM SIGOPS Oper. Syst. Rev. **35**(5), 103–116 (2001)
30. Durillo, J.J., Nae, V., Prodan, R.: Multi-objective energy-efficient workflow scheduling using list-based heuristics. Future Gener. Comput. Syst. **36**, 221–236 (2014)
31. Othman, A.B., Nicod, J.M., Philippe, L., Rehn-Sonigo, V.: Optimal energy consumption and throughput for workflow applications on distributed architectures. Sustain. Comput. Inf. Syst. **4**(1), 44–51 (2014)
32. Mezmaz, M., Melab, N., Kessaci, Y., Lee, Y.C., Talbi, E.G., Zomaya, A.Y., Tuyttens, D.: A parallel bi-objective hybrid metaheuristic for energy-aware scheduling for cloud computing systems. J. Parallel Distrib. Comput. **71**(11), 1497–1508 (2011)
33. Li, J., Liu, X., Zhao, Z., Liu, J.: Energy consumption prediction based on time-series models for CPU-intensive activities in the Cloud. In: 15th International Conference on Algorithms and Architectures for Parallel Processing (ICA3PP 2015), Zhangjiajie, China, 18–20 November 2015

CIT-Workflow: An Elastic Cloud Workflow Service

Yusheng Jiang, Jian Cao(✉), and Yan Yao

Department of Computer Science and Engineering, Shanghai Jiaotong University,
Dongchuan Roand. 800, 200240 Shanghai, China
cao-jian@cs.sjtu.edu.cn

Abstract. Workflow has been applied extensively in enterprises and
the business processes managed by workflow systems range from office
automation to production activities. Therefore, workflow systems are
becoming a common IT service for enterprises. With the development
of cloud computing, we can make use of resources provided by cloud
computing companies through a "pay-as-go" business model. Workflow
systems can also be deployed to the cloud computing environment to
offer workflow services to a large amount of users. This paper presents
an elastic cloud workflow service, which can automatically scale up and
down according to the system's needs dynamically.

Keywords: Workflow · Cloud computing · Elastic system

1 Introduction

With the development of cloud computing, more and more services are transferred from local network to cloud environment. In cloud, rescaling the server
cluster also becomes simple and cheap. The workflow, which enables the business
process to be executed automatically [1], also can make use of cloud computing
to extend it's functions. Currently, not only can a workflow model coordinate
various cloud services to achieve a complicated business logic, the workflow system can also be deployed to cloud to make use of the cloud resources to provide
workflow services to users.

By using cloud as its environment, the workflow service becomes cloud workflow service, it can take advantages of cloud computing, especially the scalability and flexibility of cloud environment. This lets the cloud workflow service be
able to easily scale up and scale down by purchasing more servers or releasing
existing servers from cloud service provider, according to the system load and
performance dynamically. There are already many studies on the self-adaptive
distributed system relying on load balancing or resource scheduling [2,3] for an
enterprise's IT infrastructure. However, in the enterprise, only it's internal fixed
resources can be used while cloud allows you lease or release resources according to your needs. Recently some research has done focusing on building such a
scalable system in cloud [4].

© Springer Science+Business Media Singapore 2016
J. Cao et al. (Eds.): PAS 2015, CCIS 602, pp. 106–113, 2016.
DOI: 10.1007/978-981-10-1019-4_9

In our previous work, we have developed a cloud workflow service [5]. The system supports manually adding and deleting servers to reconfigure the system. If servers are purchased from cloud, the administrator has to manually start and configure the server applied from cloud computing service provider before adding it to the system, and delete it from cloud after removing it from system. In this case when we want to reconfigure the system, the administrator must take care of the configuring process. This is quite inconvenient, and we hope the system is able to automatically scale up and down according to its performance. We have improved our CIT-Workflow, a workflow system which can be deployed to cloud, and to make it become an elastic cloud workflow service.

The rest of this paper is organized as follows. We describe our system's architecture in Sect. 2. Then, we introduce some critical technologies in Sect. 3. Section 4 discusses our potential future work. Finally we make a conclusion to summarize the paper.

2   System Architecture

In our previous work, we have already developed a cloud workflow system CIT-Workflow, which consists of three parts, i.e., the modeling tool, the engine service and the management platform (Fig. 1). The modeling tool provides a visualization interface to users to enable them to create their workflow models. The engine service is responsible for executing specific workflow requests, calling required services and applications to execute detailed tasks according to the workflow model. To deal with large amount of requests, the service is designed as distributed system and can be deployed to a server pool to obtain high performance and large capacity. And the management platform takes the responsibility of monitoring and controlling the server cluster. It can add and delete servers from the server pool, monitor each server's status, allocate requests, collect server's information to let administrators have a overall view about system. More detailed information about its mechanism is given in our previous paper [5].

In the previous version of CIT-Workflow, though the management platform supports rescaling the server pool, it needs the administrator to perform the

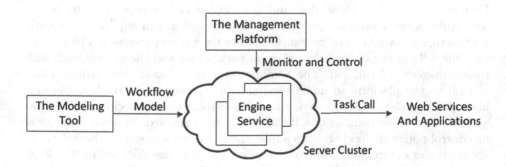

Fig. 1. The cloud workflow service structure

rescaling operations manually. This may be practicable when such operations happen rarely. But if the system needs to deal with fast changing loads, it has to rescale itself more quickly and agilely. Recently, we have moved forward to let the management platform be able to rescale itself automatically.

By using cloud computing environment such an elastic system can be implemented. For traditional IT infrastructure, we have to buy new servers to enlarge the cluster and once we bought, we can never return these servers back. But cloud servers are different, we can buy servers and return servers at any time, and we just need to pay for them according to our usage. And recently, more and more cloud service providers begin to provide APIs to users to allow programmatically accessing the cloud resources, like AWS SDK [6] and Azure SDK [7]. They provide various tools on different platforms, supporting most operations on cloud. This is the fundamental point of our new management platform.

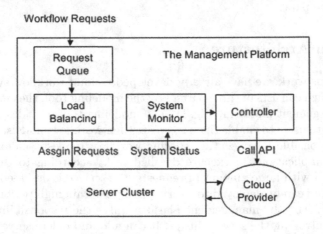

Fig. 2. The structure of the management platform

The Fig. 2 shows the structure of our new management platform. The workflow requests will first be sent to the platform. The platform maintains a request queue and pushes the requests into it. The load balancing module picks requests from the queue and assigns them to a suitable engine service in the server cluster. Each engine service is independent and can execute a whole request on its own. If no suitable servers are available yet, the requests will stay in request queue until new servers are available or previous requests have been processed. The system monitor collects the status information of each server and the system load, and records history information. The administrator can browse these status information on the platform to understand the current system status and running history. These data will also be sent to the controller to help decide whether the system needs adjustment or not, if the controller is enabled. The controller uses its control policy to decide how to adjust the server pool based on these data. If the controller determines to add or delete servers, it makes API call to the cloud provider to execute specific actions. After requests are completed and verified, the server pool gets rescaled and returns back the result to the system monitor.

3   Key Technologies

3.1   Service Deployment

Previously we use Apache Tomcat and Apache Ant to support the service remote deployment. The remote server must have preinstalled Apache Tomcat and MySQL service. Then on the management platform, we use Apache Ant to package the engine service and deploy it to remote Tomcat as an WebApp. After the deployment completes and we can access the service on the remote server, we add the server to our running server list and start assigning requests to it. This requires the server to have some preinstalled basic softwares and we must have the administrator rights to access them.

But by using cloud server, there is a much easier practice. Currently most cloud providers support users' own system images and support using these images to create cloud server. When we choose a cloud to deploy our service, first we create a server with a normal operation system and install required softwares, deploy our service on it manually, set the service to start with the system together. Then we create a system image from that server which already contains our service. Normally we do not have to pay for system images separately. We preserve the image and when we purchase a new server from cloud provider we directly use this image as our new server's system image. Thus when new server starts, the service just starts together automatically. After we get the IP of new server we can directly access the service.

Such a practice is convenient but we still have to do data sync after a new server starts up as we don't have a shared database. A shared database is convenient but it may become the performance bottleneck of the whole system. Thus we decided to use local database for each service. After a new service starts, we export the information about workflow models and other related data from database and send it to the new service. When a service receives these data, it imports these data to its local database and use them to execute a workflow.

3.2   Load Balancing

As described in Sect. 2, all workflow requests will first be sent to the management platform and pushed into the request queue, which is a FIFO queue, i.e., first coming request will be processed first. Then the load balancing module will pick requests from the queue and assign it to an engine service in the server pool.

The server's loads are measured by the number of workflow requests running on it. In our current design, each server will have an ideal requests number under which the server can achieve the highest usage ratio without obvious performance degradation. This value can be preset according to request executing history or by special experiments, or simply by experience. When we get such a preset for a server, we get the preset for all servers with the same type.

When the load balancing module picks a server to assign requests, it will first eliminate the servers which are going to be deleted soon but not yet due to unfinished requests on them. Then it searches the server list and selects the first server which hasn't reach its ideal requests load. If all servers are running under full load, the requests will be kept in request queue, until new server added or previous requests complete. Note that we do not select the server with the lowest load. Once we find server which is available we will assign requests to it, which means servers in the front of server list will always get requests assigned first. This is for the convenience of deleting servers. By this means the servers in the backend of server list can be quickly deleted when needed as they will always have the lowest load.

Most traditional load balancing policies try to assign requests equally, in that situation, it will be hard for us to find idle servers to delete, which means we will always have to wait some time for some servers to complete their current jobs and then release them. In our load balancing policy, we try to fill in servers until they reach their ideal load first, then look for other idle servers. This policy enables most requests centralised on some servers, so that we can remove idle servers flexibly when system is under a low load.

3.3 Control Policy

The main goal of an elastic system is to rescale the system for processing different level system loads, improving the performance and reducing the cost. The control policy directly influences the effect of the rescaling process. It determines when and how the system add or remove servers.

Different control policies can have quite different effects. It is a core part of an elastic system. Good policies can even predict the request change and do system adjustment in advance. In our system we use arriving request number and current system load as our main criteria. When request number increases or system is under a high load pressure, we will add new servers to enlarge the server pool, so that more requests can be processed in time. When request number decreases or keeps stable, i.e., the system has a low load level, we will remove servers to reduce the cost. Besides, because most cloud servers are paid in a hourly base and starting a new server requires some extra time, the control policy will try to avoid frequent adjustment. When request number is fluctuating, we take conservative actions, adding servers quickly to guarantee the service quality but deleting servers slowly to avoid fluctuation of servers.

Specific control policy can be a research topic and we have described our policy in another paper in detail. In brief, we use time series prediction to predict future requests number. Then divide request coming patterns into three patterns, increasing, decreasing and stabilize/fluctuating and apply different control policy in different pattern. In increasing mode we add servers in advance to avoid the delay time when activating a new server; in decreasing mode we remove servers when system is in low load and add servers conservatively; in stabilize/flucuating mode, we try to stabilize the server number while keeping the system's performance.

4 Future Work

The system is still in developing. The Fig. 3 shows a screenshot of our management platform's interface.

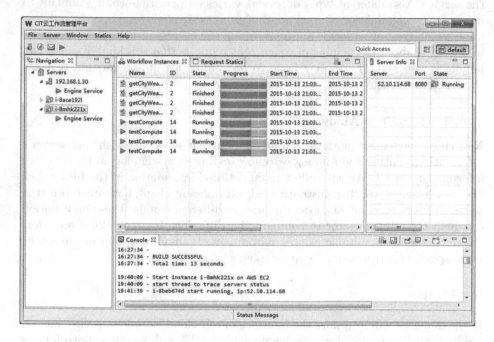

Fig. 3. The management platform of CIT cloud workflow service

We are still working to improve the system. Here are some function points to be added:

4.1 Integrate with Multiple Cloud Environments

As the server pool is deployed on cloud, the platform has to use API or other tools from cloud providers to execute various cloud actions. However there is no standard way to send requests to each cloud provider. Some cloud providers provide SDK in Java or other languages, some only support HTTP requests, and also some don't. Besides, different cloud provider supports different actions. We have to manually read the help doc from cloud providers, add support for each cloud provider and try to provide similar functions. Currently we only support AWS and Aliyun. We plan to support more cloud providers in the future. As different providers may provide resources with different performance and price, the administrator is able to choose a suitable cloud provider or multiple services to host the workflow service.

4.2 Improve Controller's Function

Currently our control policy uses current request number and ideal request number to determine the system load level. However the ideal request number is hard to determine as different workflow models may produce different loads to the service. Also different types of servers varies on performance, it's difficult to measure the capabilities of every kind of servers. It is also important to determine which type of servers is appropriate for the current situation. Besides, the information about the number of concurrent requests that this kind of server can handle is also helpful. We may hope to use online learning algorithm to learn these information by itself.

4.3 Performance Analytics

Now the management platform can provide current server list with each server's information, waiting and running workflow instances, system log and some other information via UI. We will collect more detailed information in the future, like how much time a workflow instance needs on different cloud, how much time the server needs to start up, and even the price of different clouds. These information can help the administrator to compare the difference between cloud providers and pick the most suitable cloud environment. It may even help on designing more specific control policy to apply to special environment.

5 Demo

In the demo we will introduce the interface of our cloud workflow service, including its modeling tool and management platform. We will record a video for the demo. First we will demonstrate the basic functions of our system. Then we demonstrate how the system handle requests and rescale itself during several hours. And shows its static data collected in the process.

6 Conclusion

In this paper we introduced our elastic cloud workflow service. It is extended from our original cloud workflow service. The system deploys engine service on a server pool to support high capacity and expandability. And we use cloud as its environment to support rescaling. The management platform of the system can automatically rescale the system on cloud based on the system's load.It is able to assure service quality without wasting the computing resources when load changes. In the future we will keep developing the system to improve its performance and usability.

Acknowledgement. This work is partially supported by China National Science Foundation (Granted Number 61272438,61472253), Research Funds of Science and Technology Commission of Shanghai Municipality (Granted Number 15411952502, 12511502704).

References

1. Van Der Aalst, W., Van Hee, K.: Workflow Management: Models, Methods, and Systems. MIT press, Cambridge (2004)
2. Heinis, T., Pautasso, C.: Automatic configuration of an autonomic controller. In: ICAC 2008. International Conference on an Experimental Study with Zero-configuration Policies. Autonomic Computing, pp. 67–76. IEEE (2008)
3. Gillmann, M., Weissenfels, J., Weikum, G., Kraiss, A.: Performance and availability assessment for the configuration of distributed workflow management systems. In: Zaniolo, C., Grust, T., Scholl, M.H., Lockemann, P.C. (eds.) EDBT 2000. LNCS, vol. 1777, pp. 183–201. Springer, Heidelberg (2000)
4. Li, H., Venugopal, S.: Using reinforcement learning for controlling an elastic web application hosting platform. In: ICAC, pp. 205–208 (2011)
5. Jiang, Y., Cao, J.: CIT workflow - a workflow platform faced to integration of cloud service. In: CBPM (2014)
6. Tools for Amazon Web Services. https://aws.amazon.com/tools/
7. Download Azure SDKs and Tools—Azure. http://azure.microsoft.com/en-us/downloads/

Integration of Process Management and IT on Cloud

Jinzhu Liu[✉], Fuhua Han, and Fei Liu

ActionSoft Co., Ltd., Beijing 100085, China
{jackliu,liuf}@actionsoft.com.cn

Abstract. In this paper, four common problems encountered by enterprises in business process management are summarized based on our practical experience in the field of business process management for more than ten years. AWS BPM related products and AWS BPM implementation methodology are presented, and we also demonstrate a complete solution for integrating process management and IT (Information Technology) on cloud computing through real client cases.

Keywords: AWS · BPM · PaaS · Process management · Process analysis · Process execution

1 Introduction

Process management concept was introduced to China since the late 1990s, and enterprise has different cognitions and practices on process management at different development stages. Enterprise management has experienced initial management based on personal experience, function management, management of local process by information-based means and current stage of promoting overall continuous improvement by comprehensive process management [1]. Enterprises constantly improve their recognition on BPM. More and more enterprise managers are planning to start BPM related projects in the aspect of demand. Enterprises are also increasing investment since they have implemented BPM projects. Operation mode of traditional enterprises is facing unprecedented challenges with advent of Internet thinking.

Three major technological revolutions of software industry are reviewed. The first revolution occurred in 1960, which realized separation of database from application logic, and it was not necessary for application software development to focus on underlying storage structure. Professional database vendors and DBMS system were produced, thereby greatly improving development efficiency of application software. The second revolution refers to visual programming technology produced in 1984. GUI-separation of user interface and application logic was realized. Application software development speed was greatly accelerated. The third technology innovation was produced since 2000, and business process management was separated from application logic, which was represented by generation of BPM product actually. Different requirements are proposed to enterprise management ability at development stages of organization. We discover that management software is changing from function-driven operation to process-driven operation [2] (Fig. 1).

© Springer Science+Business Media Singapore 2016
J. Cao et al. (Eds.): PAS 2015, CCIS 602, pp. 114–119, 2016.
DOI: 10.1007/978-981-10-1019-4_10

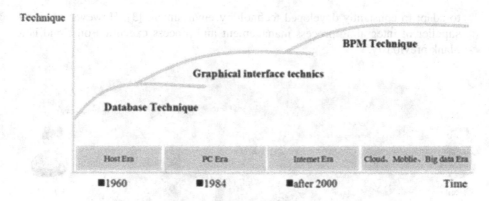

Fig. 1. Three technology separations of management software industry

2 Problems in Process Management

It is discovered that enterprises generally do not have tools and methods about process plan and process analysis according to our practice in process management field for more than ten years. In such a case, process analysis results cannot be implemented after a lot of fund is invested. Figure 2 shows the disconnection of enterprise value chain which causes many problems in enterprise process management as follows.

1. Business process is available, and there is no process management. Process carrier refers to a large number of paper documents, which are scattered throughout the enterprise without uniform management. Process description, optimization and change are not effectively controlled and managed, thereby leading to great gap among documents or between document and actual business. Since process manual and management documents are electronic, huge workload of adjustment and modification is produced, and enterprise internal management can not be effectively evaluated and analyzed.

2. Many process and management systems are seriously separated. Uniform specification is insufficient among departments, management systems and processes of systems. Since the focuses are different, different business processes are produced aiming at realization of the same business under various management systems. Some processes are mutually conflictive and hence it is difficult to fully exert comprehensive benefits of management systems.

3. There is no uniform execution system between process management and IT. Management and system can not be linked. Analysis results are not implemented by the aid of information system, or difference is produced. The process execution process is not transparent. The process has insufficient process compliance and risk control ability. There is no uniform process monitoring system. Process monitoring dimension is not comprehensive. Analysis on process execution performance is one-sided.

4. In addition, with the emergence of Internet, cloud computing and other novel technologies, enterprises implementing BPM system urgently demand relevant solutions

to adapt to constantly developed technology environment [3]. However, solution supplier of integrating process management and process execution on cloud is a blank previously.

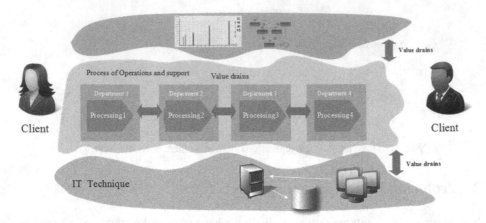

Fig. 2. The disconnection of the enterprise value chain

3   Integration of Process Management and IT

Internal integration of process management system is realized by comprehensive process management of ActionSoft AWS BPM (as shown in Fig. 3), thereby promoting linkage between process management and system coordination. Centralized integration of information system is supported. The plan is composed of AWS CoE excellence process center (as shown in Fig. 4) and AWS BPM platform.

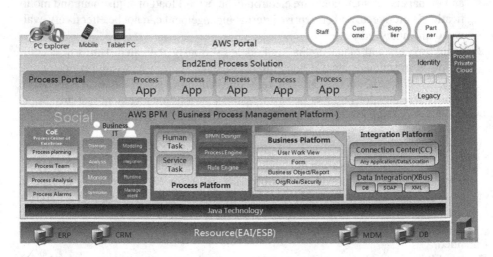

Fig. 3. Technical framework of AWS BPM

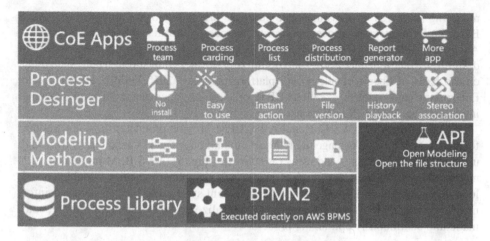

Fig. 4. Technical framework of AWS CoE

AWS CoE assists enterprise to form strategic advantage centered on process ability, which can promote enterprises to evolve to process organization centered on client. Overall management ability of enterprises can be improved. Many formal (or virtual) inter-departmental process teams can be effectively organized by using AWS CoE. Respectively responsible process areas can be decomposed by social network groups. Consistent BPM tool is adopted for establishing process framework and unifying

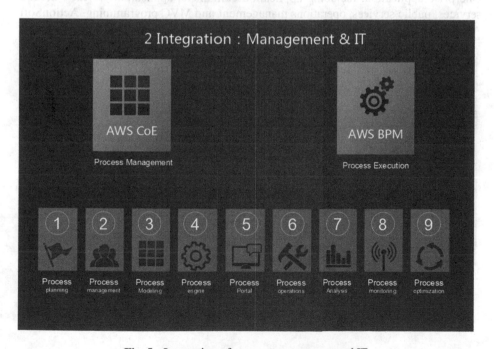

Fig. 5. Integration of process management and IT

modeling standards. Process management tasks are continuously implemented. AWS BPM platform integrates process platform, business platform and integration platform. Advanced BPM and SOA architectures are adopted. Enterprises can be assisted for designing, executing, monitoring and optimizing business flow through pure Web mode on the basis of open technology and platform such as J2EE, Eclipse, HTML 5, etc. Enterprises can establish excellent BPM process management architecture through implementing AWS BPM solutions.

Two platforms of AWS CoE and AWS BPM provide product component support for all stages of process management. Process management and IT are integrated through using unified technologies and ideas (as shown in Fig. 5). AWS BPM implementation methodology is adopted, and product platform is combined to achieve double closed-loop management of business process management system and business process information system, thereby assisting enterprises to realize explicit, systematic, curing, automatic and continuously optimizing process.

4   Cloud BPM Technology

ActionSoft launched AWS PaaS (Platform as a Service) in 2015 with emergence of Internet, cloud computing and other emerging technology, and it is a professional PaaS for the field of enterprises, which contains all features of BPM (Business Process Management) technology. As shown in Fig. 6, AWS PaaS provides complete platform service ability in order to achieve the delivery, operation and maintenance objectives of enterprise application, including application containers, application services, mobile services, public services, operations management and MVC programming. ActionSoft AWS PaaS fills the blank of cloud process management in China.

Fig. 6. Technical framework of AWS PaaS product

5   Case Background of Field Demonstration

In this demonstration, the case of integrating process management and IT of one 3C company is displayed through AWS PaaS cloud platform. The case client is one of the largest mobile phone chain enterprises in China. Its chain stores are distributed in more than 20 provinces and more than 300 cities all over the country, and the company is provided with 2000 stores, more than 300 after-sales service outlets and more than 20000 employees. The stores are distributed all over the country with rapid expansion and development of the enterprise. Conditions of market at all levels are different, thereby leading to too long management chain. However to implement group strategy, management system and operational process rapidly and accurately, how to make the whole operation line management free from disorder due to industry feature of personnel mobility, and how to correspond to enterprise business change, organization change and management change due to market change, and smoothly implement strategy adjustment have become urgent problems.

A management platform integrating enterprise finance, purchase, inventory and sales, manpower, logistics, after-sales service, business process, administrative management, etc. is created for operation and management chain of enterprise through group business process management system established for one year based on AWS BPM platform. Management features of chain retail industry are combined are combined. Trichotomy is adopted for establishing a uniform process system with clear process, clear responsibility and powerful monitoring according to categories of 'strategy flow', 'business flow' and 'support flow'. Standardized management strategy of 'unifying multiple stores into one pattern' is guaranteed. Enterprises can be assisted for establishing and strengthening management concept oriented by process. Store business execution efficiency and specification can be effectively improved, data analysis is provided for optimizing continuous business flow. Leaders are assisted for making a decision through mastering accurate and real-time information at any time. More than 300 main processes cover most business areas, which plays an important role in the enterprise.

References

1. Han, W., Zhang, H.: Configurable modeling technology of business process modeling notation. Comput. Integr. Manuf. Syst. **19**(8), 1928–1934 (2013)
2. Fang, R.: Design and realization of whole process modeling tool based on BPMN/BPEL. Master thesis, Beijing University of Posts and Telecommunications (2015)
3. Wang, Z.: Design and realization of business process on PaaS platform. Master thesis, Xi'an University of Electronic Science and Technology (2011)

Workflow Management System for Numerical Weather Prediction

Weifeng Wang[1(✉)], Wen Zhang[2], Qunbo Huang[1], Tingfang Wang[1], and Kaijun Ren[2]

[1] Meteorological Centre of Air Force, Beijing 100843, People's Republic of China
wwfeng990@163.com
[2] Academy of Ocean Science and Engineering, National University of Defense Technology,
Changsha 410073, Hunan, People's Republic of China
renkaijun@nudt.edu.cn

Abstract. Numerical weather prediction system is a complicated software system which generally involves thousands of computational processes which must be controlled and operated automatically by computers for the efficiency and correctness. This paper designs and implements a workflow management system. With this system, a new kind of workflow process definition language is adopted and the corresponding parser is accordingly developed. On the base of the workflow process definition file, a workflow execution engine is deployed to automatically execute all defined business processes or jobs. During job execution, The monitoring commands which were inserted into the job shell script files in advance can be used to collect related information which also can be displayed on the monitor interface. Moreover, checkpoint strategies are employed to quickly recovery system after it crashes abnormally. Finally, a workflow demo is taken to demonstrate the management functions and operating rules of the system.

Keywords: Workflow management system · Workflow process definition language · Monitoring commands · Checkpoint

1 Introduction

The numerical weather prediction (NWP) [1] business system generally consists of global and regional four-dimensional variation data assimilation, medium-range prediction for global aviation meteorological elements, short-term numerical prediction for limited-area aviation meteorological elements and other sub-systems. There are complex relationships among these sub-systems, and the execution of all jobs must conform to time sequences which are controlled by a set of shell script files. In the past, a lot of time must be spent to modify the parameters used in a job ahead of the execution of NWP system and all scripts must be started manually one by one. In addition, during the running of jobs, many types of unpredictable errors often occur for example the hardware and software problems of supercomputers so that the running status of jobs must be monitored and errors must be corrected timely which certainly will bring huge challenges if monitored and corrected only by labor force. As such, a workflow system is necessary to automatically control the execution of all defined processes or jobs and

© Springer Science+Business Media Singapore 2016
J. Cao et al. (Eds.): PAS 2015, CCIS 602, pp. 120–127, 2016.
DOI: 10.1007/978-981-10-1019-4_11

help monitor some errors simultaneously. Workflow Management System (WfMS) [2, 3] transforms the real-world business processes into a form of computerized representation, and provides the functions of execution and management. This paper designs and implements a workflow management system for NWP. System architecture and some key technologies with the system are introduced in the following sections.

2 System Architecture

Figure 1 illustrates the major components of the system architecture. The architecture shows that the workflow management system for numerical weather prediction consists of three main programs including management monitoring tool, client-side tool and workflow engine. They are respectively discussed in the following sections.

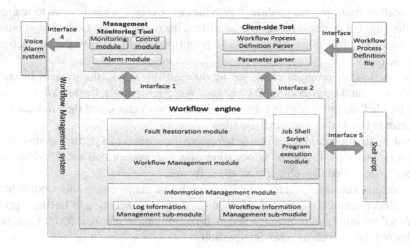

Fig. 1. System architecture

2.1 Management Monitoring Tool

The management monitoring tool is a program for man-machine interface, whose function is to monitor and control the health of business process. It comprises monitoring module, control module and alarm module. The functions of each component are as follows:

- Monitoring module: it collects real-time execution information of business processes, and the information is presented to the clerks in the form of text, tables, drawings and other forms in the man-machine interface.
- Control module: it provides an interface as well as operating rules for the clerks, puts the behavior of the clerks into the specific stream and then sends the stream to the workflow engine.
- Alarm module: it implements the judgments for the business execution exception conditions and the invoking of the voice alarm system to notify the clerks.

2.2 Client-Side Tool

The client-side tool is designed as a communication tool to realize the interaction between participants (clerks and job shell script files) and the workflow engine. In addition, according to the system technical characteristics, the process definition parser is included in this main program.

2.3 Workflow Engine

The workflow engine is the core program, providing a variety of management functions, which includes workflow management module, information management module, job shell script program execution module and fault restoration module. The functions of each component are as follows:

1. Workflow management module: it provides the execution mechanism to realize the automation of business process. In additions, it can control the execution of business process and manage workflow process definition table to support the interactions of participants.
2. Job shell script program execution module: in this paper, the process instance is called job, and a job has a shell script file. With this module, the workflow variables in the shell script files were replaced with relevant values, and then the shell script parser is invoked to execute these files.
3. Fault restoration module: it is able to restore the system to a particular state with no loss or less loss of critical information after the system crashes. Obviously, that greatly improved the reliability of the software.
4. Information management module: log information management sub-module records the clerks' actions and related information about the execution of business process. Workflow information management sub-module manages shell script files, output files of jobs, as well as other information.

3 Key Technology

3.1 Workflow Process Definition Language

The workflow process definition contains all necessary information about the business process for sure that it can be executed by the workflow engine. Necessary information previously mentioned involves the information for navigation between jobs and rules, job shell script program that would be invoked, input and output variables, etc. The workflow process definition language is designed to meet the demands of the workflow process definition. Particularly, it is unambiguous semantics that it can be interpreted by computers. Some main elements of the language are described as follows:

1. Structure Elements: the structure of the organization for the business process is described by three elements, such as suite, family and task. A task, that is a job, has a shell script file whose name is the same as the task. A family is used to collect tasks together or to group other families. A suite is a collection of families, variables,

repeat elements. One workflow process definition file has only one suite. A numerical weather prediction business system can be divided into several parts according to the function, each part contains some jobs. NWP business system can be described by the following program codes.

```
suite NWP; family 6H_YHPROC; family 00;
task start_yh4dvar_job; endfamily; endfamily; endsuite;
```

2. Dependence Elements: the dependence can be defined by time, trigger, date and other elements. Some elements, such as time and data, indicate the conditions of starting time. The trigger element can define event-dependence, status-dependence, etc.

```
task start_yh4dvar_job; time 13:00; task
ftp_to_get_teee; trigger start_yh4dvar_job == complete;
```

The codes show that the start_yh4dvar_job will be executed at 13:00. The ftp_to_get_teee will be executed after the start_yh4dvar_job is complete.

3. Event Elements: the event of the job is defined by two elements, and they are event and meter. The purpose of an event is to signal partial completion of a job. Meter is an extension of the event. In some tasks there may be many events which are set in order, e.g. an event might be set every six hours, more than 40 events. The event elements can be able to trigger another job which is waiting the partial completion of the job. This is described by the following program codes.

```
task start_yh4dvar_job;event myy; task ftp_to_get_teee;
trigger start_yh4dvar_job:myy == set;
```

4. Circulation Elements: circulation structure in the business process can be defined by two elements, such as repeat and cron. The codes show that the NWP will be executed every day.

```
suite NWP; repeat day 1; endsuite;
```

5. Variable Element: the edit element defines input and output data in the workflow process.

```
task ftp_to_get_teee; edit ANAL_DATE '20100508';
```

The codes show that the ANAL_DATE is defined as time variable which will be used in the ftp_to_get_teee.

3.2 Monitoring Mechanism

When the job is executed, the system is supposed to monitor the status of the job. The system provides monitoring commands to gather status information for the job. The monitoring commands include active, aborted, complete, event, meter and so on. The active, aborted and complete commands indicate the execution status of the job. The event and meter commands can monitor event status. These commands are inserted into the job shell script files according to relevant rules. When a command is executed, the system will update the monitor status table via the message bus. Figure 2 shows the detailed process of the monitoring mechanism.

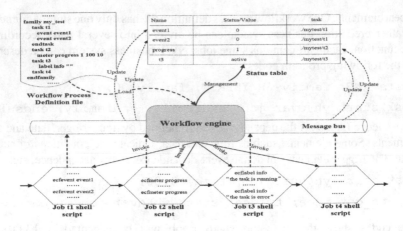

Fig. 2. Monitoring mechanism

3.3 Workflow Execution Mechanism

In this paper, the workflow execution mechanism is a technique which the system uses to automatically run the job in accordance with the rules of the workflow process definition. A job will not start running, until all dependencies are satisfied as well as reach a particular status.

Status reflects the state the job is in. The Status includes UNKNOWN, QUEUED, SUSPENDED, SUBMITTED, ACTIVE, ABORTED, COMPLETE. Figure 3 shows the transition process among status for a job. These are described as follows:

1. When the workflow process definition file is loaded into the workflow engine and the clerks do not start the business process, the status of the job is UNKNOWN.
2. When the begin command is executed, the status of the job is QUEUED. In addition, the status of the job which has circulation elements is QUEUED after the complete command in the job shell script file is executed.
3. When the starting conditions of a job are satisfied and the system is resolving job shell script, the status of the job is SUBMITTED.
4. After the active command in the job shell script file is executed, the status of the job is ACTIVE.
5. After the aborted command in the job shell script file is executed, the status of the job is ABORTED.
6. After the complete command in the job shell script file is executed, the status of the job is COMPLETE.
7. After the suspend command is executed, the status of the job is SUSPENDED.

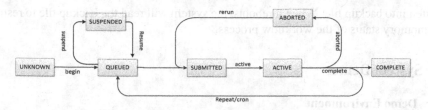

Fig. 3. Transition process among status for a job

A job may have event-dependence, time-dependence and other dependence, there are many uncertainties before the starting conditions of the job are satisfied. Therefore, the system uses a timer to judge starting conditions of all jobs by the way of polling. In addition, all the jobs whose starting conditions are satisfied will be executed in parallel. Figure 4 shows the details that how the system finds and operates executable jobs. Task, family and suite are all regarded as node in the figure.

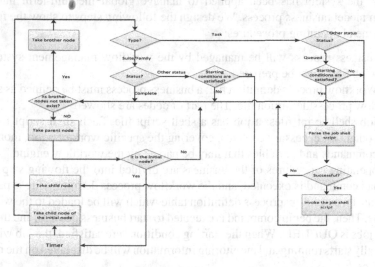

Fig. 4. Finding and operating process of the executable jobs

3.4 Fault-Tolerance Mechanism

When the system crashes, it can recover to a normal state without loss or less loss of workflow process information. Therefore, the clerks are only required to rerun a job instead of rerunning the entire business process, when the system is rebooted. Checkpoint is a snapshot of the state for the system at the end of a transaction and at the beginning of another transaction [5]. Referring to the checkpoint backup technology, the system backup workflow process information every one minute using a timer, and only stores the latest and the second new backup file. Boost serialization technology is used to solve the issues of information backup and recovery. Serialization Technology transforms the workflow process memory status to a binary stream format, which is written into backup file. When it reboots, the system will read the backup file to restore the memory status of the workflow process.

4 System Demo

4.1 Demo Environment

The system is developed with C++ language based on the boost library and motif library, and the job script uses shell script language. Therefore, Demo environment is made up of the following parts: RedHat Linux 64-bit operating system, boost library, motif library and shell script interpreter.

4.2 Demo Steps

Currently, the system has been applied to manage global medium-term numerical prediction model business process. We design the following steps to show the function and the general operating procedures:

1. If a business process will be managed by the workflow management system, the following files must be prepared:
   (a) Workflow process definition file: a business process must be defined as a workflow process definition file. The part of codes are showed
   (b) Job shell script files: a job has a shell script file. Each shell script file must contain all necessary contents, covering the specific work of a job, monitoring commands and variables that may be parsed by the workflow engine.
2. The operating procedures of the business are divided into the flowing steps. First, the load command is executed, and the workflow process definition file is parsed to generate the workflow process definition table which will be loaded to the workflow engine. Then, the begin command is executed to start business process, and the status of all jobs is QUEUED. When the starting conditions are satisfied, the job will automatically starts running and monitoring information will be displayed on the monitor interface.
3. In some cases, we have no choice but to start the job manually. Under the circumstance, the execute command can be executed to force the job executed. In addition, the delete command can be executed to remove all the dependencies of a job, and the job will start running after a while.
4. When the jobs are running, once abnormal operations occurs, they are supposed to be detected in time and monitoring tool can notify the clerks via voice.

Figure 5 shows the components of the business process, the status of the job and some user operating buttons. In the figure, QUEUED is set to light blue, COMPLETE is set to yellow, and ABORTED is set to red.

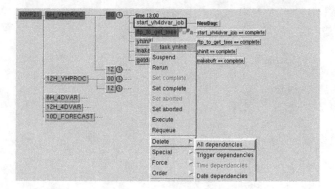

Fig. 5. Monitoring view of the business process (Color figure online)

5 Conclusion and Future Work

The paper described a workflow management system applied to numerical weather prediction business system. The introduced system can automatically run business processes and provide powerful monitoring capabilities, which can greatly reduce the workload of the people. What's more, the efficiency of running NWP system has been improved. Nevertheless, there is still a lack of graphical process description tool with the system, and we are trying to develop graphical interface definition tool referring to the process definition language.

Acknowledgement. This work is supported by National Natural Science Foundation of China (Grant No. 61572510, 61502511) and China National Special Fund for Public Welfare (Grant No. GYHY201306003).

References

1. Haltiner, G.J.: Numerical Weather Prediction. Wiley, New York (1971)
2. Lawrence, P.: Workflow Handbook 1997. Workflow Management Coalition. Wiley, New York (1997)
3. Elnozahy, E.N., Alvisi, L., Wang, Y., Johnson, D.B.: A survey of rollback-recovery protocols in message-passing systems. ACM Comput. Surv. **34**(3), 375–408 (2002)

Author Index

Printed in the United States
By Bookmasters